Diesel Common Rail and Advanced Fuel Injection Systems

Philip J.G. Dingle
Ming-Chia D. Lai

SAE International
Warrendale, PA

All rights reserved. No part of this publication may be reproduced, stored in a retrieval system, or transmitted, in any form or by any means, electronic, mechanical, photocopying, recording, or otherwise, without the prior written permission of SAE.

For permission and licensing requests, contact:

SAE Permissions
400 Commonwealth Drive
Warrendale, PA 15096-0001 USA
E-mail: permissions@sae.org
Tel: 724-772-4028
Fax: 724-772-4891

SAE Global Mobility Database®
All SAE papers, standards, and selected books are abstracted and indexed in the Global Mobility Database.

For multiple print copies, contact:

SAE Customer Service
E-mail: CustomerService@sae.org
Tel: 877-606-7323 (inside USA and Canada)
724-776-4970 (outside USA)
Fax: 724-776-1615

ISBN 0-7680-1257-0

Library of Congress Control Number: 2005927223

Copyright © 2005 SAE International

Positions and opinions expressed in this book are entirely those of the authors and do not necessarily represent those of the organizations to which the authors are affiliated or SAE. The authors are solely responsible for the content of this book.

SAE Order No. T-117

Printed in the United States of America.

Other SAE titles of interest:

The Diesel Engine
By Daniel J. Holt
(Order No. PT-109)

Diesel Engine Reference Book
By Bernard Challen and Rodica Baranescu
(Order No. R-183)

Practical Diesel-Engine Combustion Analysis
By Bertrand D. Hsu
(Order No. R-327)

**Diesel Particulate Emissions
Landmark Research 1994–2001**
By John H. Johnson
(Order No. PT-86)

**Diesel Nitrogen Oxide Emissions
Landmark Research 1995–2001**
By John H. Johnson
(Order No. PT-89)

For more information or to order a book, contact SAE at
400 Commonwealth Drive, Warrendale, PA 15096-0001;
phone (724) 776-4970; fax (724) 776-0790;
e-mail CustomerService@sae.org;
website http://store.sae.org.

Contents

Preface 7

Executive Summary 9

Chapter One
    History and Background 15

Chapter Two
    Injector Nozzle and Spray Technologies and Their
    Impact on Emissions 25

Chapter Three
    Common Rail Fuel Injection Equipment 43

Chapter Four
    Hydraulic Electronic Unit Injectors 75

Chapter Five
    Unit Injectors and Unit Pumps 81

Chapter Six
    Current Application Issues 91

Chapter Seven
    Future Outlook and Technology Trends 99

References 127

Acronyms and Definitions 133

About the Authors 137

Preface

It is a strange dichotomy, but true, that while in the heavy-duty sector North American manufacturers arguably lead the world with their diesel truck engine technology, to the extent that the engines are emulated and sold across the globe, in the diesel passenger car sector North America lags behind the world. To some extent, this is due to an unhappy experience with passenger car diesel engines in the early 1980s, but there are also several socioeconomic reasons for this situation. As a result, North America today is a strong bastion of spark ignition (SI) hegemony, a situation that might easily be expected to continue well into the future. However, the times are changing. Proposed CO_2 reduction legislation ostensibly favors the diesel engine, and the forward-thinking product planners can see that a gasoline/diesel mix in the fleet can open a new niche and bring flexibility to the Corporate Average Fuel Economy (CAFE) calculations.

Elsewhere in the world, diesel engine penetration into the light-duty vehicle market has advanced rapidly over the past twenty years and particularly over the last five years. In large measure, this recent growth spurt is due to the availability of the so-called "common rail" fuel injection system, which has almost single-handedly transformed the diesel engine, making it seriously competitive with the gasoline engine as a private vehicle powerplant. Now, a number of factors are converging that improve the prospect for the availability of diesel engines in the North American light-duty market, not the least of which is the market shift toward sport utility vehicles (SUVs), for which diesel engine characteristics are most admirably suited. These characteristics include fuel economy, invincible low-speed torque, and all-around good drivability enabled in part by sophisticated engine subsystems such as the fuel injection system and variable geometry turbochargers (VGTs).

However, if passenger car diesels are to return to North America in any volume, either through the import of engines produced elsewhere or particularly in the case of new engines designed and assembled locally, it is necessary to get all aspects of the process right this time. No one wants a repeat of the dismal experience of the early 1980s. Many factors are different this time. There is a greater understanding of diesel combustion now within the global diesel community, extensive experience with light-duty engines is available from Europe, and precision in manufacture has improved. However, by the same token, the performance and emissions challenges have intensified, too. Some things have not changed. The engineers in product development and manufacturing who will be charged with specifying, optimizing, and producing these engines will perforce be steeped in the lore and language of SI engines and may not be attuned to the subtle but important differences with the compression ignition (CI) engine. If that were not the case, there would be no need for this book.

Here, then, in fairly general terms, is a resource that addresses the important aspects relating to the diesel fuel injection system, explaining how we have arrived where we are, what systems are available today, providing pointers for what aspects are important and what aspects are not, and finally looking at the current state of development and projecting the likely technology path for the future. There is no doubt that diesel development requires a slightly different mindset, and this book should help in establishing that. The book is not aimed exclusively at either the light-duty or heavy-duty sector, but it points out where those sectors diverge in their solutions to the common problems.

Executive Summary

In the 110-year history of the diesel engine and its associated fuel system, there have been three watershed periods in which technological progress was particularly rapid. The first and most spectacular occurred in the 1920s, when "air-blast" injection gave way to "solid" injection. A less spectacular revolution took place 35 years later in the mid-1950s and into the 1960s when the rotary distributor pump came onto the market. Finally, the introduction of the modern electronically controlled common rail injection system in the late 1990s has significantly changed the outlook for the diesel engine. In each case, technological advances made to the injection system opened huge new market opportunities that previously had not been accessible. In the case of solid injection, it brought the diesel engine into the automotive truck market. Rotary distributor pumps opened the light-duty automotive market for indirect injection (IDI) diesels, and common rail has enabled the emissions-controlled direct injection (DI) diesel engine to be successful in several market sectors. Most visible of these is the light-duty automotive market, particularly in Western Europe, where diesel penetration is currently at an aggregate average of 44% and is still climbing. But the impact of common rail technology also can be felt at the other end of the spectrum. Modern cruise ships all now have common rail injection systems, so that they may avoid the visible smoke emissions that were an unwelcome attribute over much of the propeller-load curve with the previous traditional fuel injection equipment (FIE).

Essentially, common rail has established a new benchmark for diesel injection systems that most traditional systems are not able to assail, and as a result, they are falling by the wayside. In-line and rotary pump-line-nozzle (PLN) systems are in rapid decline, and only the unit injector family is in a position to compete. Although not a new concept, common rail, which gets its name from the high-pressure accumulator typically running down one side of the cylinder head and feeding all injectors from this common source, has logically moved most of the fundamental system functions close to the point of injection. Traditional PLN systems controlled fuel metering, injection timing, and pressure generation at the pump, with only the injector being responsible for atomization. On the other hand, common rail integrates fuel metering, injection timing, and fuel dispersion all within the injector, which imparts a huge advantage in precision of control relative to previous systems. Moreover, with the pump freed of these responsibilities, it is now able to modulate the system pressure as a means of controlling overall injection rate, which again is a valuable feature not previously available.

Drivers for the ready acceptance of the new electronically controlled common rail systems are largely emissions related. On one hand, legislated reductions in tailpipe emissions demanded continuous FIE improvements with respect to parameters such as injection pressure, injection rate flexibility, and control precision, which had to be met if current diesel engine families were to remain in production. On another front, the common rail system has been found to be ideal for the light-duty DI automotive engine as noted above, where its sophistication of control over the combustion process has effectively made this engine the equal and in some respects superior to the previously hegemonic spark ignition (SI) engine. In turn, the popularity of the light-duty DI diesel is due in part to the desire for economical mobility in an era of rising motor fuel taxation and a desire to limit CO_2 greenhouse gas emissions.

In comparison to the gasoline engine, the light-duty diesel shows to advantage in the areas of specific fuel consumption, low-speed torque, and fun-to-drive factor. While improvements are being made continuously to the diesel engine, OEMs are still obliged to focus on nitrogen oxide (NOx) and particulate matter (PM) emissions, cold starting and associated idle noise, and initial powertrain cost. In all of these, the FIE plays a decisive role. Indeed, the FIE is and always has been key to the diesel combustion process, and although exhaust aftertreatment is under intensive development, the usual decision is to exploit the FIE capabilities to their fullest extent. The cost-to-benefit ratio almost always favors minimizing engine-out emissions in the first instance rather than expecting aftertreatment to clean up a polluting engine that has less capable FIE. Additionally, FIE sophistication usually brings side benefits such as reduced noise, vibration, and harshness (NVH) for which the consumer is prepared to pay, whereas exhaust aftertreatment provides few consumer-perceived benefits and a likely fuel consumption penalty.

The technical challenges associated with continuous emissions improvement in all classes of engine, as well as attainment of powertrain cost and refinement goals, drive the solutions, which may vary from one market sector to another. At present, common rail has made only modest inroads into the heavy-duty truck market where the unit injector is now the dominant fuel system. There are multiple reasons for this situation, covered in the appropriate chapter here. However, a reason that will become increasingly important in the future is that the unit injector offers unique and compelling benefits of its own that are not yet matched by current common rail systems. These include significantly higher injection pressure capability, always beneficial where high levels of diluent are involved, and in the latest two-valve electronic unit injector (EUI) a capability for multiple injections with shot-to-shot pressure and rate control. Nevertheless, looking to the future, it can be assumed that both common rail and unit injectors will strive to converge on the "ideal" fuel injection system specification, and in so doing, they will each appropriate the superior features of the other system. The present paradigm, however, views common rail as offering greater refinement and unit injector as offering higher specific power ratings by virtue of the higher-pressure capability—hence, the dominance of common rail in light-duty applications and EUI in heavy-duty applications, with hydraulic electronic unit injection (HEUI) systems making a strong showing in medium-duty applications.

Fuel injection technology road maps provide a pointer for the development direction of diesel engines as a whole, because many of the advances made to the engine are dependent on FIE technologies. Generation-by-generation increases in peak injection pressure capability have been driven by the demand for higher specific power outputs and lower PM emissions for a given power level. Pilot injection capability has been an ideal for decades, but only now with common rail has it worked well enough that it can be fully exploited for reduction of noise and NOx. To a significant extent, control of combustion heat release has been enabled by multiple injection capabilities that, in effect, "digitize" the injection rate profile. Advances in manufacturing precision and software control algorithms, both in the manufacturing plant and in the engine control unit (ECU), have greatly improved unit-to-unit consistency and refinement. All of these features and many more to come are changing the perception of diesel engines from one with negative connotations to one that has a rapidly growing acceptance and often preference in ownership.

The positive features of diesel engines for automotive applications notwithstanding, many open questions remain to be answered as we look to the future. Diesel engines compare unfavorably to SI engines on the basis of initial cost, and this situation will be aggravated by the significant costs of exhaust aftertreatment. Moreover, the active nature of the aftertreatment regeneration requires regular dosing with a reductant, frequently the fuel itself, which detracts from the overall vehicle fuel economy. These factors affect the market appeal of the diesel engine, leaving the door open for SI engine improvements that could be decisive in market share retention. However, in the struggle for market share, technological improvements appear set to continue at a rate at least as rapid as in the recent past, with piezoelectric control actuators and variable orifice nozzles being examples of FIE progress that will enable new low-emissions combustion opportunities. While these technologies would appear to add to the initial system cost, they will be perceived as highly cost effective if they enable homogeneous charge compression ignition (HCCI), mixed-mode, or other low-temperature combustion strategies that minimize the exhaust gas aftertreatment requirements. Other opportunities for cost reduction will come from intensive efforts at system integration by, for instance, exploiting the significant synergies between the fuel system and engine valve actuation systems.

If one considers that the diesel engine requires twenty parts of air to consume one part of fuel by weight, then it will be apparent that the air-handling system has a very important role to play. This book explains the advances that have been made recently and some that will be made on the FIE side of this equation. However, one can anticipate similar advances soon will be made on the air-handling side, in the form of flexible valve event functionality such as variable valve actuation (VVA). Versions of this technology are available today on SI engines, but it is neither so readily applied to compression ignition (CI) engines nor is it so obviously advantageous for conventional diesel combustion. However, when any of the low-temperature combustion modes such as HCCI or premixed charge compression ignition (PCCI) are contemplated, then the combination of flexible FIE and flexible air handling will be a key enabler and will bring with it numerous performance enhancements that will only further improve the competitive position of the diesel versus the gasoline engine.

This eventuality, however, will force further changes to the FIE because the logical progression for VVA valvetrain is from mechanical systems to lost-motion systems to "camless" systems. At this point, the cam-driven EUI is likely to give way to an intensified FIE system, at the same time as the common rail system moves to an intensified system, too, because sustained pressures higher than 2000 bar are challenging to generate efficiently in a single stage. It appears that FIE is likely to converge on a common solution that, at this time, is projected to be an intensified injector with direct needle control, offering variable nozzle flow areas with perhaps selectable spray angles, injection possible across the full 360-degree crank-angle domain, shot-to-shot injection pressure control, and likely embodying active injection rate control.

Of course, all this assumes that the current diesel engine paradigm continues unchallenged. It only requires someone to develop a low-emissions high-swirl combustion system that delivers competitive fuel consumption, and the race for super-high-injection pressures could be over. Likewise, HCCI could be a game-changing technology that alters the diesel engine industry radically. Under the current paradigm, diesel engines are relatively low-speed, high brake mean effective pressure (BMEP), high-torque, relatively heavy powerplants. High BMEP does not seem possible under true HCCI operation, so an engine concept designed explicitly for HCCI, in which multiple moderate BMEP power strokes are delivered at high frequency in a lightweight structure, could revolutionize the industry and the FIE requirements, too. Many HCCI research engines operate with low-pressure port or gasoline direct injection (GDI) fuel systems, not expensive high-pressure injection. This change in engine paradigm could pose a threat to the current internal combustion engine (ICE) industry because it opens the door to new entrants.

Chapter One

History and Background

In the mid-1890s, when the Franco-German engineer Rudolf Diesel (1858–1913) was developing his compression ignition (CI) internal combustion engine (ICE) concept, the subsystem that took the longest time to perfect was the fuel injection system. He discovered that although it was relatively easy to make the engine run with the most basic of fuel supply system, a strong relationship existed between the sophistication of the fuel injection system and engine performance in terms of power, efficiency, smoothness, and controllability. This relationship has proved to be fundamental, because it still holds true today, 110 years later.

Consider for a moment the role of the fuel injection system as applied to a CI engine. In the most fundamental terms, it is required to meter a desired quantity of fuel, to introduce that fuel to the combustion chamber at the desired crank-angle timing, and to arrange for the intimate mixing of that fuel charge with the compressed air in that chamber. These three basic requirements sound simple to achieve, and indeed Rudolf Diesel managed to get his engine to fire initially in 1894. However, it took another three years of concentrated work on the fuel and combustion systems before the engine was deemed to operate well enough for the concept to be licensed.

If we now look in more detail at some of the issues that make the difference between a basic fuel injection system and one that is fully satisfactory in a demanding application, an understanding of the problems will be gained.

Metering

The greater the thermal efficiency of an engine, the smaller is the amount of fuel required to deliver a given power output. Rudolf Diesel's first engine recorded a thermal efficiency of 26.2%, approximately on par with a modern spark ignition (SI) engine. Today, it is possible to obtain in excess of 50% from the CI cycle. This implies that the fuel injection system must be capable of consistent and accurate metering of very small quantities of fuel. Because injection pumps traditionally have been of the positive displacement type, it has been convenient to measure and express the delivery in volumetric terms, such as 50 mm^3/injection. Today, it is more common to use the gravimetric term of milligrams per injection (mg/injection). For instance, 50 mm^3/injection resolves to 50 × 0.83 (typical fuel density) = 41.5 mg/injection.

Timing

The point in time at which the fuel is introduced to the combustion chamber has a strong relationship to many operating parameters of the engine, not the least of which is thermal efficiency. Again, the accuracy and precision of the time at which injection commences relative to the desired crank-angle position must be of a high order, with ±0.25 crank degree being regularly obtained today. Achieving this level of control with start-of-injection timing flexibility over the engine speed-and-load map with a high-pressure hydrodynamic process has traditionally been challenging.

Dispersion

Considering now the presentation of the fuel to the compressed air, the objective has always been to achieve a high utilization of the air in the chamber under full-load conditions. High air utilization translates directly into high specific power output when all other factors remain constant. This requirement has led to many ingenious designs of combustion chamber and fuel atomizer combinations over the years, and this remains an area of active development both now and, it can be projected, into the future. The two main variables that can be controlled are the air motion within the chamber during the combustion event, and the fuel spray pattern and momentum. High air utilization requires a known and well-controlled air motion, in conjunction with an accurately targeted spray plume pattern. Typical air motions are swirl and/or squish, which may be induced by a combination of inlet conditions and piston or chamber geometries. Conventional fuel dispersion techniques have been either air-assisted atomization or high-pressure liquid-only injection. In Rudolf Diesel's day, air utilizations were probably on the order of 50%; today, values of 90% are achievable. An air utilization of 100% would mean that the bulk cylinder contents had burned at $\lambda = 1$, or stoichiometry.

Airless Injection

The fuel injection system that Rudolf Diesel developed for his original engine was of the air-assisted or "air-blast" atomizer type, in which compressed air is used to disperse the fuel into the combustion chamber. Therefore, all the original licensees of Diesel's CI engine initially utilized this atomizing technology. But because it perforce limited the maximum engine speed to approximately 500 rev/min and thus low specific output, there was an incentive to develop a new fuel injection system that could eliminate this impediment. Nevertheless, air-blast injection remained the only viable fuel system until superseded during the late 1920s by liquid-only or "solid" injection.

The exigencies of World War I demanded rapid improvements in all technologies that could be used as instruments of war, and military desires for greater engine power, but with less bulk and weight, spurred engine innovation and development. Even before the war, the British naval contractor Vickers was developing liquid-only or "airless" injection systems (including a common rail fuel system) that dispensed with the air compressor and associated hardware [Ref. 1.1]. This line of development offered the promise of meeting the military objectives.

But at this time, the CI engine was still quite unsuited to road transport applications. However, the attributes of splendid low-speed torque and the highest available thermal efficiency as offered by the cycle were highly valuable to commercial vehicle operators. The size of this potential market was the incentive necessary to encourage new fuel system development.

The engineer generally credited with developing the first commercial airless injection system and associated combustion system was Prosper L'Orange (1876–1939). Initially, he worked on the problem of elimination of the air-blast injection system with its expensive compressor while at Deutz, but he was lured away to focus on this issue at Benz & Cie in 1909. Also in 1909, L'Orange invented a precombustion chamber of the type in which the fuel charge is injected and combustion initiates under favorable conditions, before it expands as it burns, through a connecting passage into the main cylinder. Divided-chamber geometries such as this are often referred to today as indirect injection (IDI) systems. In a reflection of Rudolf Diesel's difficulties in developing his injection system, the gestation of an airless injection system was a protracted undertaking for L'Orange, too. Thus, it was not until 1920 that Benz began the production of a prechamber CI engine with an airless injection system for agricultural applications, and in 1924, the first road-going truck was offered for sale.

The Birth of the Fuel Injection Equipment Industry

This work was the key that hugely extended the market potential of the CI engine and led to the birth of the fuel injection equipment (FIE) industry. At one stroke, Diesel's engine was shown to be a candidate for road transport applications, and designers were no longer shackled by the air-blast injection system. They were free to invent their own injection system and combustion chambers, and they did. Two models emerged. In the first, which was a continuation of current practice, the engine manufacturer designed and built his own fuel injection system. In the second, specialist suppliers emerged who developed proprietary designs that were available for application to customers' engines. Initially, during the 1920s, the first model prevailed, due in part to inertia and the lack of alternatives. Starting in the late 1920s and growing rapidly during the 1930s, the second model began to dominate, particularly in Europe. It must be apparent that for an engine builder skilled in the design and manufacturing art of the five "C's" (i.e., crankcase, crankshaft, cylinder head, camshaft, and connecting rod), provision of a piston air compressor for the air-blast injection system was well within his capabilities. But when it came to the airless injection system, the necessary manufacturing skills were quite different, with the emphasis being on small, high-precision hardened and ground components. This dichotomy provided an opportunity for the growth of the fuel injection system supplier industry, and it is no accident that the suppliers who seized this opportunity were, in many cases, the same companies that supplied high-class precision equipment such as magnetos to the SI engine builders. They could see that the growth of this new engine concept would imperil their profitable magneto business. In turn, the availability of proprietary FIE enabled many SI engine builders to add diesel engines to their portfolio without a commensurate investment in tooling or even, in many cases, a deep understanding of diesel combustion.

The Pump-Line-Nozzle System

The fuel injection system developed by Prosper L'Orange was of the type that has become known as pump-pipe-nozzle, or alternatively, pump-line-nozzle (PLN). As suggested by this appellation, there are three conspicuous components: an engine-driven high-pressure "jerk" pump, a pipe to conduct the fuel pulse to the injector, and an injector that incorporates an atomizing nozzle (Figure 1.1).

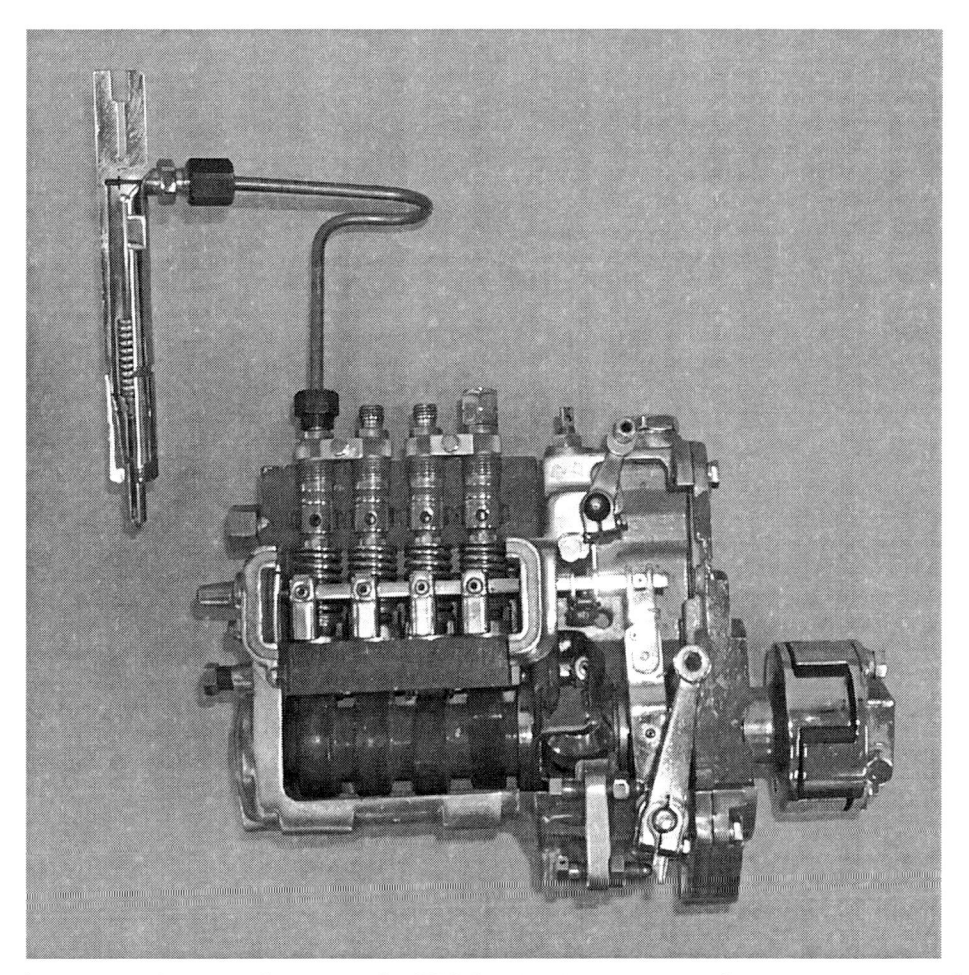

Figure 1.1 A pump-line-nozzle (PLN) injection system, showing a sectioned pump, high-pressure pipe, and spring-controlled injector.

This basic architecture quickly became the convention because it packaged conveniently on most engines, and it worked demonstrably well. Another likely reason is that there are many possible variations of detail design for the components that would deliver the required functionality, so the first company to market with the concept could not patent and protect against all possible permutations. Until the advent of distributor pumps, this combination of PLN components was and today still is duplicated for each engine cylinder.

Rotary Distributor Pumps

The second truly profound change to occur in the history of diesel FIE was the development of the rotary distributor pump during the 1950s. Replacing one high-precision pumping and metering assembly per cylinder with a single pumping module plus a distribution rotor created a smaller, cheaper, lighter injection system than the foregoing in-line pumps (Figure 1.2). This development opened the small high-speed light-duty passenger car and industrial engine market to the diesel engine. The most commercially successful of these pumps was the internal cam ring design invented by Vernon Roosa (d. 1989) and licensed to Lucas CAV, which went on to produce some 50 million during the following 50 years.

Figure 1.2 Rotary distributor pump (sectioned), line, and pintle injector.

By contrast, heavy-duty truck and industrial engines were beyond the range of most distributor pumps and so were specified with large-frame-sized in-line pumps, or predominantly in the United States with unit injectors (see Chapter Five). In this latter arrangement, both the injector nozzle and the pumping mechanism are unified in one housing and of necessity are located in the cylinder head. Thus, the high-pressure pipe and its associated volume are eliminated, and the resultant improvement in hydraulic stiffness aids both precision of fuel control and efficient pressure generation. Pumping plunger actuation is by overhead camshaft and rocker mechanism or, on occasion, with the addition of a stiff pushrod if the cam is mounted high in the cylinder block. From this market background, the common rail fuel system emerged in the mid-1990s to establish new standards of performance and functionality. In so doing, it changed the landscape of diesel fuel injection, causing the demise of in-line and rotary pumps and spurring the further development of unit injectors, as illustrated by Figure 1.3.

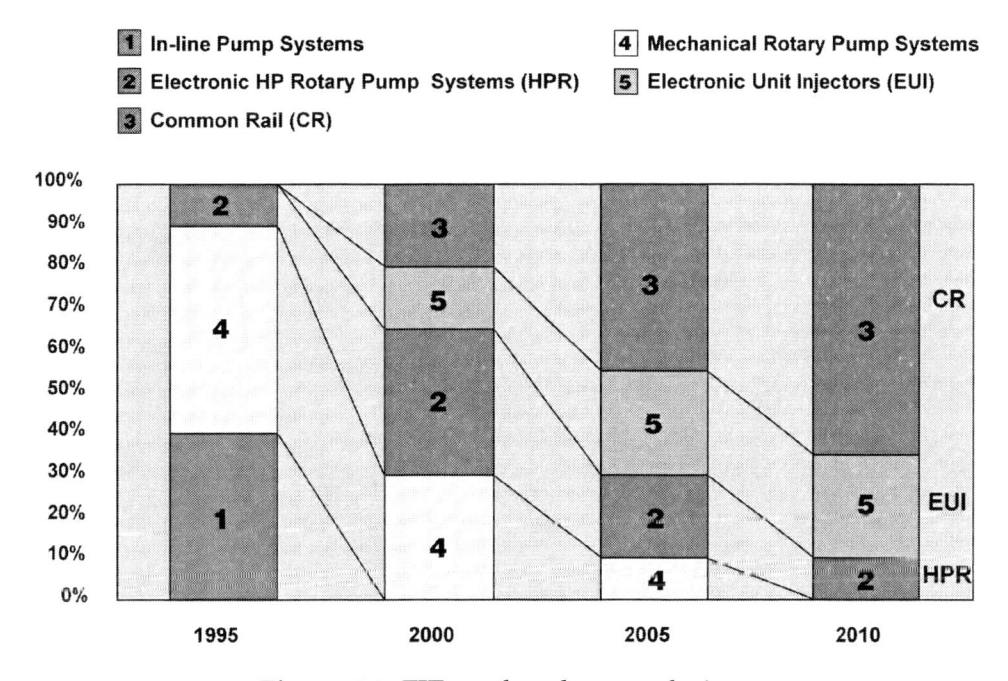

Figure 1.3 FIE market share evolution.

Combustion Systems

With the Otto (SI) engine, a homogeneous fuel and air mixture is drawn into the cylinder, compressed, and, if we may neglect any quenched remnants, upon ignition the entire combustion space contents are consumed. However, this is not the case with the CI engine. A key requirement in the design of a CI combustion system is to arrange the chamber geometry so that at top dead center (TDC), the maximum amount of air possible is accessible to the injected fuel, in a chamber having the lowest surface area attainable. Moreover, if air motion is to be a factor in the combustion process as it usually is, it must be orderly and expected. Random turbulence simply will not do; the flow patterns must be anticipated so that the air and injected fuel can be arranged to cooperate, if maximum air utilization and minimum emissions are to be obtained. These requirements force the designer to go to great lengths to minimize top ring crevice volume, piston crown to fire-deck clearance, and valve pocket recess depth, because at TDC, these combined volumes can represent a large percentage of the total air. By way of example, Richards *et al.* in Ref. 1.2 indicates that in the case of a heavy-duty truck direct injection (DI) engine, this fugitive volume can amount to some 40%. Although in reality, much of that air does come out of hiding early in the expansion stroke to play a minor role in the combustion process, it remains a worthy goal to avoid the dead pockets. Given these constraints, many combustion systems are theoretically possible, and, in turn, many concepts have been reduced to practice over the years with varying degrees of success, as illustrated in Figure 1.4 [Ref. 1.3].

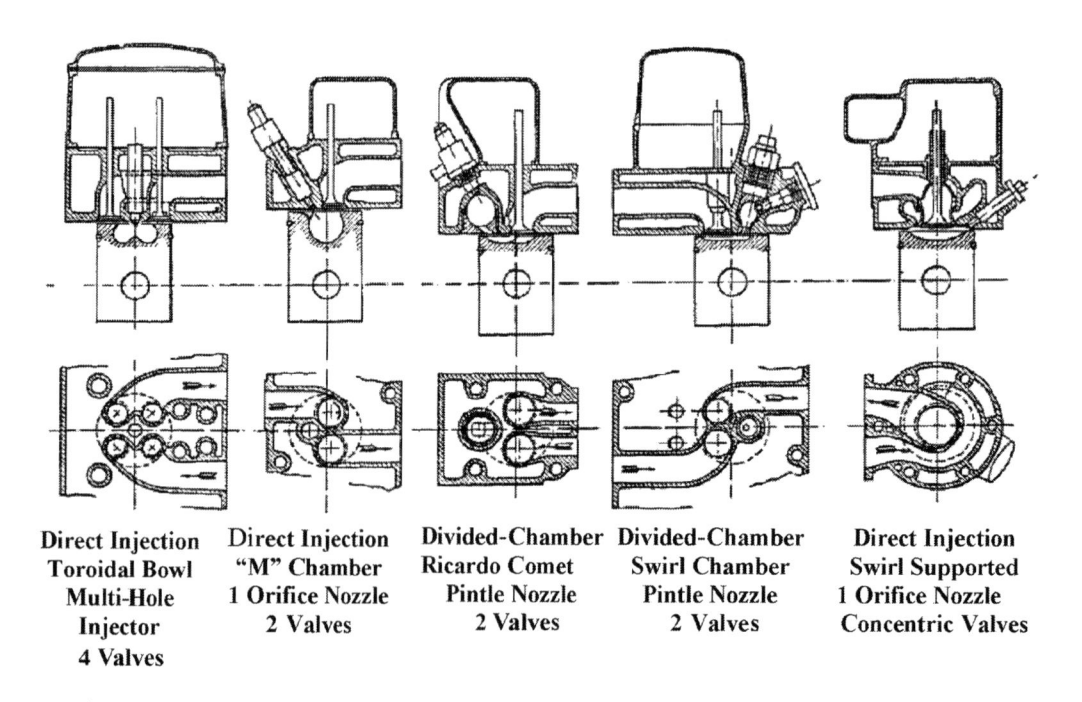

Figure 1.4 Illustration of various historical combustion systems [Ref. 1.3].

A balance exists between the combustion chamber designer's ideas and the fuel system supplier's manufacturing capabilities, and it is necessary that they work closely together as a team without the concerns of one taking precedence over the other. Indeed, only by having a thorough and deep appreciation of the other's constraints can they be expected to deliver a competitive combustion system. A paramount and defining aspect of conventional diesel combustion is that the fuel and air mixing take place over approximately 35 crank degrees around TDC, and in that short time, all the injected fuel must find and mix with sufficient air to locally reach a flammable equivalence ratio. To make this happen while meeting the performance and emissions targets requires all the skills and ingenuity of these two team members.

Manufacturing Requirements

The peak injection pressure requirements of a successful diesel combustion system, even the less demanding divided-chamber designs, demand extremely tight clearances between the fuel system mating surfaces, particularly in the case of the pumping plunger and its sleeve. These diametral clearances may be on the order of 0.003 to 0.006 mm and are necessary to minimize leakage because the pressures are beyond the capability of conventional seals. Indeed, seals would add undesirable compliance and dead volume to the system. At ambient temperatures, diesel fuel has a dynamic viscosity of approximately 3.0 centistokes, and, being a Newtonian fluid, that drops to around 1.0 centistoke at 100°C. Given these values, the need for extremely tight running clearances becomes apparent. Furthermore, it can be appreciated that maintenance of tight clearances over extended plunger lengths, as is necessary to seal the high pressures, requires similarly high standards of geometric straightness and roundness if the plunger is not to stick in its bore. When a close-clearance plunger is associated with a conical valve seat, as is the case with a conventional inwardly opening nozzle needle (Figure 1.5), extremely high standards of concentricity must be achieved between the valve and the seat if acceptable levels of seat tightness are to be provided.

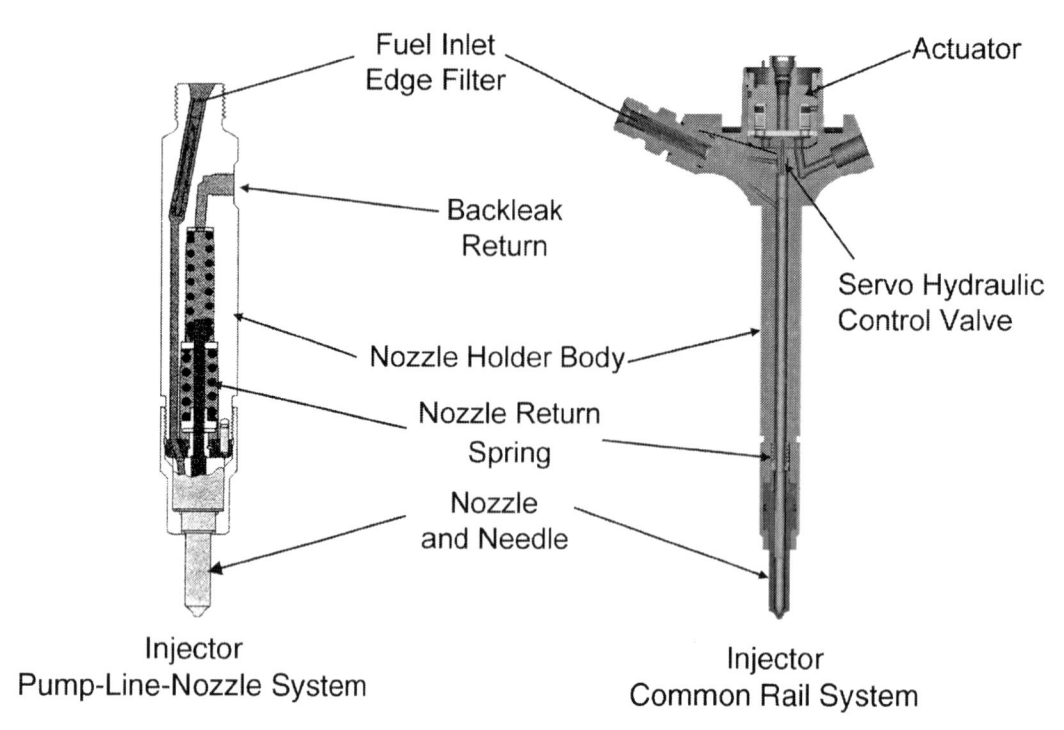

Figure 1.5 Injector terminology: (left) conventional, and (right) common rail.

Therefore, the independent FIE manufacturer has taken on the task of designing, developing, and manufacturing a device that is capable of close integration with engines of many different types, and of providing the functionality required of the engine in the wide variety of applications to which such engines are deployed. An important aspect is that the fuel system should endeavor to provide some feature or functionality that delivers a competitive advantage or differentiation for that engine in the marketplace. Moreover, this application flexibility and strength of attribute must be provided at a cost that is competitive against those vertically integrated engine manufacturers, where the engine and fuel system are designed as a whole. As with most engine accessories, however closely integrated, the FIE also must be designed to have a service life that is of the same order as the host engine. Because of the high cylinder pressures inherent to the concept, it is usual for CI engines to be of rugged construction relative to the typical Otto engine; therefore, they generally offer very high durability. This characteristic must be matched by the FIE, and, in turn, this has a profound effect on the manufacturing operations. Thus, the manufacture of diesel FIE has historically proved to be an expensive and capital-intensive proposition.

Chapter Two

ctor Nozzle and Spray
chnologies and Their
mpact on Emissions

consideration of the fundamentals of fuel injection equip-
), as outlined earlier in the first chapter, will cause one to
the conclusion that the injector nozzle in whatever form it
the key component of the system. Successful diesel opera-
mploying air blast, pressure, or other atomizer technology,
in the case of direct injection (DI) combustion systems—
quality of fuel atomization and dispersion. Because this func-
of the nozzle, one could argue that the future success of the
meeting the emissions reduction challenge will depend in
on technological advancements made to this device.

the technology trends in the nozzle design and the spray
tour of the compression ignition (CI) engine combustion and
emissions processes is required. Diesel combustion is essentially "mixing controlled," which means that the rate of heat release is largely a function of the efficiency with which fuel droplets can be evaporated and mixed with the air, so that on a local level, they are brought close to stoichiometry and thus into the flammable range. Factors that enhance this process include higher injection pressures, a more rapid air motion, and higher air temperatures. This process takes a finite time, however, and particularly so in the case of the first parcels of fuel to enter the combustion chamber, because this is when the pressures and temperatures are at their lowest values. Thus, there is a distinct time delay between the start of injection and the start of combustion. This variable, which depends on engine temperatures, effective compression ratio, and fuel cetane number, must be taken into account during engine calibration.

In traditional single-injection diesel combustion, fuel is being continuously introduced to the combustion chamber throughout the delay period, so that once ignition has been established by one of the early fuel parcels, pressures, temperatures, and reaction rates rise rapidly as all the fuel then in the chamber is consumed. This relatively uncontrolled combustion, generally referred to as the premixed phase, results in high rates of pressure rise and is largely responsible for the undesirable but characteristic diesel knock, and it contributes to NOx formation. By appropriately tailoring the injection rate profile or by employing pilot injection, one can smooth out the intensity of the initial bulk auto-ignition. These measures are in common use today. At low-idle and off-idle conditions, significant and clearly audible reductions are possible when pilot injection is invoked, in comparison to a traditional main-only injection. Further, the flexibility of modern fuel systems allows this reduction to be maintained under varying ambient temperature, coolant temperature, and exhaust gas recirculation (EGR) calibration conditions.

In comparison to spark ignition (SI) engines, CI engines produce much less unburned hydrocarbon (UBHC) and carbon monoxide (CO), due to the higher combustion temperature and abundance of oxygen in the combustion chamber (fuel-lean operation); however, CI engines produce more particulate matter (PM, black sooty smoke or white smoke at cold start, consisting of dry carbonaceous soot and unburned volatiles, soluble organic fraction, or SOF) and nitrogen oxides (NOx). The pollutant formation process in diesel combustion, which in reality is quite complicated, may be visualized in the combustion picture shown in Figure 2.1. Soot is primarily formed in the fuel-rich mixing zone, while NOx is formed in the hotter reaction zones at the edge of the fuel plume. The darker soot-clouds and the brighter flame front surrounding them are made more visible by forcing the plume to be impinged and squeezed through a small-crevice skimmer in this instance.

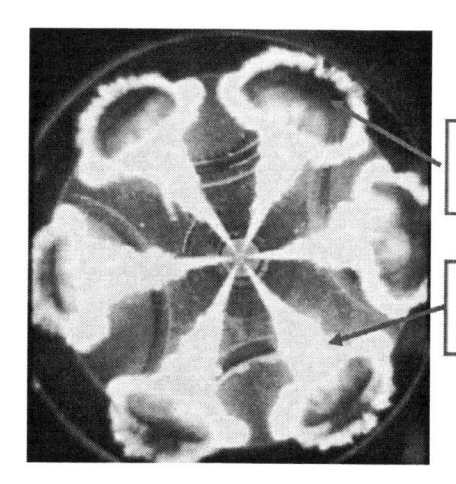

Figure 2.1 Visualization of combustion in a quiescent combustion chamber at full load typical of a heavy-duty engine. The spray plume has been forced into a quartz crevice, revealing the flame and soot formation structure.

High injection pressure can provide better spray atomization and air-fuel mixing, and it has been recognized as a key to high combustion efficiency and a low-smoke combustion process in the conventional diesel combustion mode. However, increased NOx emissions are associated with such a strategy. The conventional strategy for reduction of engine-out NOx, in addition to the more traditional injection timing retardation, is to use EGR to reduce local reaction temperatures and the subsequent thermal NOx formation. However, increasing EGR has the adverse effects of increasing PM emissions due to the reduced oxygen concentrations, hence pushing the requirement for higher injection pressure. This soot-NOx trade-off and the in-cylinder pollutant control strategies are shown in the Figure 2.2.

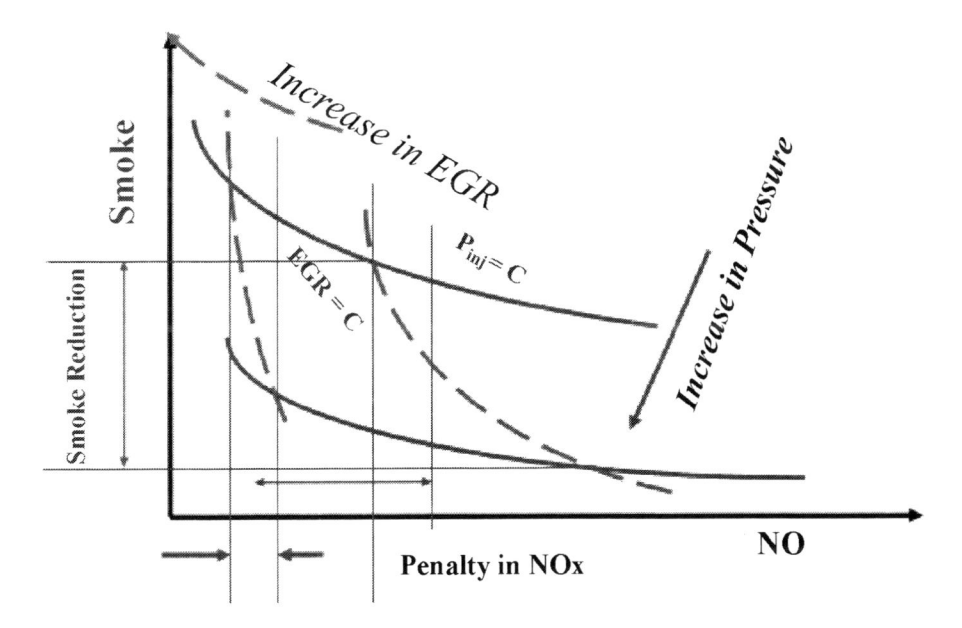

Figure 2.2 Soot and NOx trade-off and reduction strategies.

At the end of injection, control is largely lost again because the last parcels of fuel go in at low pressure with low mixing momentum and are often poorly atomized. Under unfavorable conditions of air motion, cylinder pressure, and temperature, these final fuel elements can result in smoke, as the rate of fuel-air mixing slows down and the hydrocarbons (HC) are "cooked" into PM before they have a chance to reach the flammable range. Responses to minimize smoke at this point include rapid injection termination through fast needle seating, and efforts to maintain adequate air motion even as the piston is descending. Reducing the sac volume or utilizing a valve-covered orifice (VCO) nozzle minimizes the amount of fuel exposed to the combustion environment and reduces UBHC or PM emissions. A close high-pressure post-injection has been found helpful in some situations because it can reenergize the end of combustion. Efforts to reduce NOx production through injection timing retard can, if overdone, lead to end-of-injection fuel parcels impinging on the lip of the reentrant piston bowl, resulting to slow rates of mixing and hence smoke (Figure 2.3).

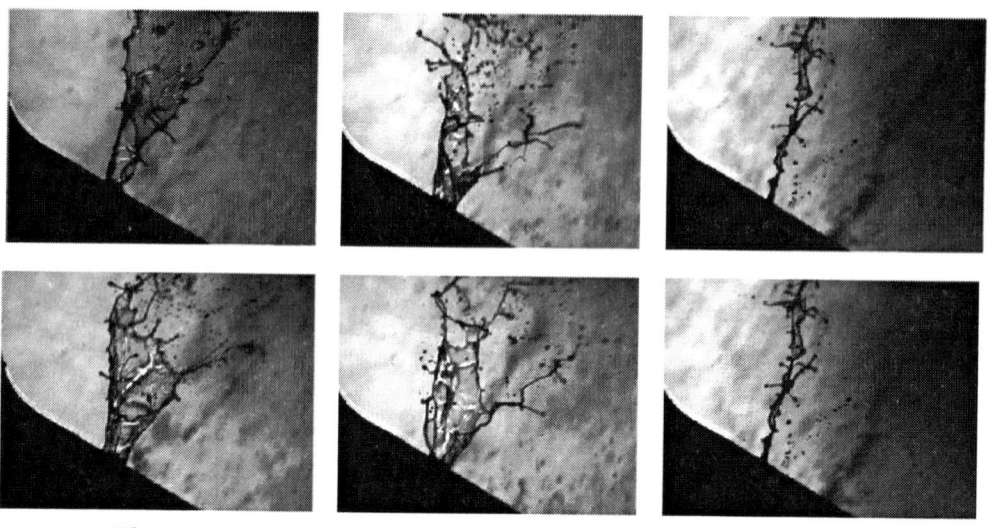

Figure 2.3 The microscopic visualization of diesel spray from a VCO nozzle at the end of injection, the breakup of which has greater effect on emissions than at the start of injection.

There is significant difference between the heavy-duty and light-duty diesel engines other than the size of the engine bore. The heavy-duty engines usually enjoy a relatively quiescent and open chamber design to minimize pumping loss, while counting on the high injection pressure alone to take care of the fuel-air mixing. However, light-duty engines have much smaller bowl diameters that border on the limit of the liquid-phase penetration of the spray, and they usually require swirl and specific bowl geometries to help the fuel-air mixing. Figure 2.4 shows the spray-bowl interaction visualization of a light-duty engine under pressurized nitrogen ambient condition at a density comparable to that at top dead center (TDC). The reentrant shape of the bowl is not only designed to interact with the charge swirl motion to generate more turbulence, but also to direct the fuel plume around the bowl to maximize air utilization. Therefore, the nozzle protrusion and included spray cone angle of the injector, as well as the shape of the bowl and the swirl ratio of the charge air motion, are all important factors in the engine combustion design.

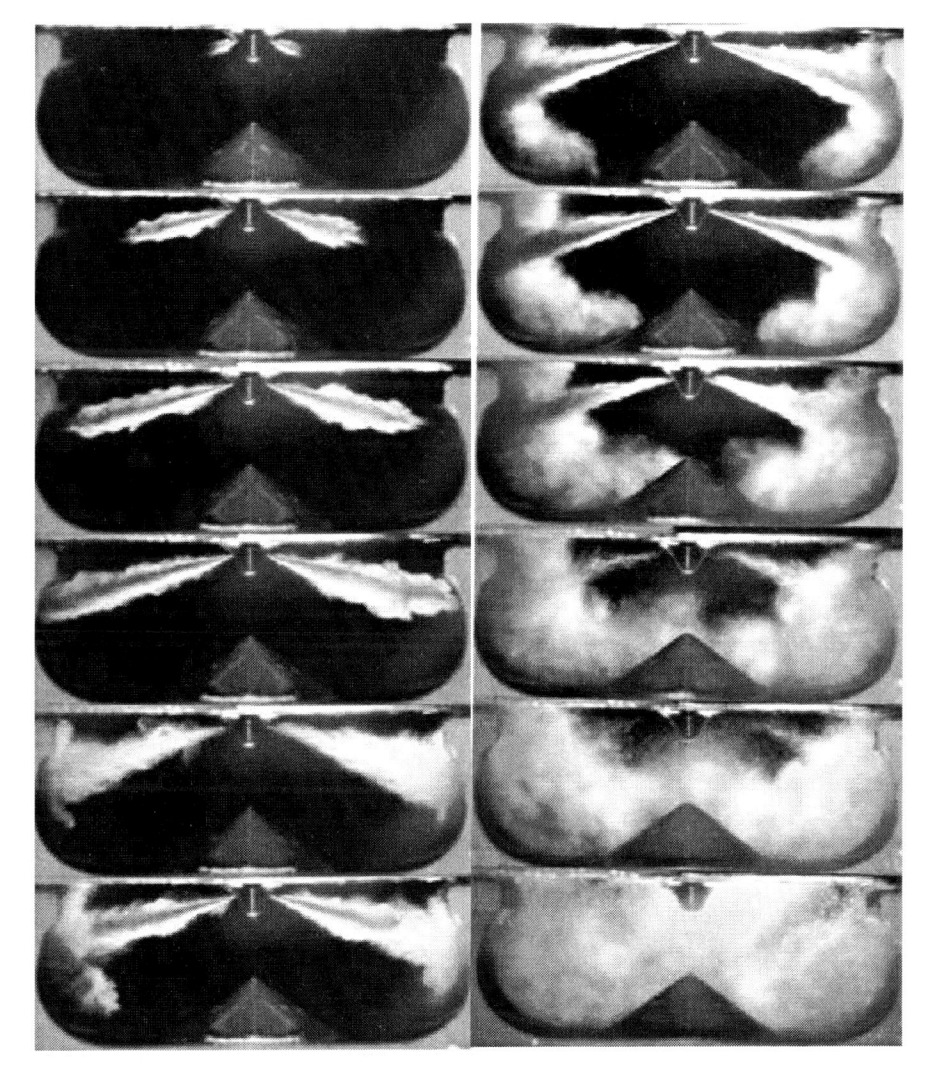

Figure 2.4 Spray-bowl interactions of a light-duty engine combustion system.

Injection Rate Profile and Injection Strategies

Optimizing the injection rate profile has been recognized as a key factor in controlling the temporal development of the fuel spray formation and the subsequent combustion and pollutant formation processes. The injection rate profile is defined as the plotted characteristic of the fuel quantity injected into the combustion chamber as a function of time or crank-angle degree. Technology development has converged on a boot-shaped injection rate, in which a low initial rising rate favors low NOx emissions and combustion noise, a later high injection pressure to burn out the soot, and a quick turnoff to minimize the adverse effects of the end of injection on UBHC and PM.

Recently, with better understanding of the mixture formation process, injection timing control has been integrated with air motion control (i.e., swirl, tumble, and squish) and EGR control to simultaneously improve the combustion and emissions processes. For example, combining a late injection strategy, heavy EGR, and high swirling mixing flow, a lower-temperature more premixed combustion phase occurring later in the expansion process can significantly reduce the soot and NOx simultaneously. Among the champions of this concept is the Nissan modulated kinetic (MK) combustion process [Ref. 2.1]. Alternatively, Figure 2.5 shows that combining an early first injection to prepare a lean premixed charge reaction, followed by a later second injection to support the diffusion combustion phase under low-temperature, low-O_2 conditions, has also been shown to provide simultaneous reduction of NOx, PM, and fuel consumption. This "uniform bulky" combustion system (UNIBUS) is championed by Toyota [Ref. 2.2].

Figure 2.5 Basic injection concept for the Toyota UNIBUS combustion system.

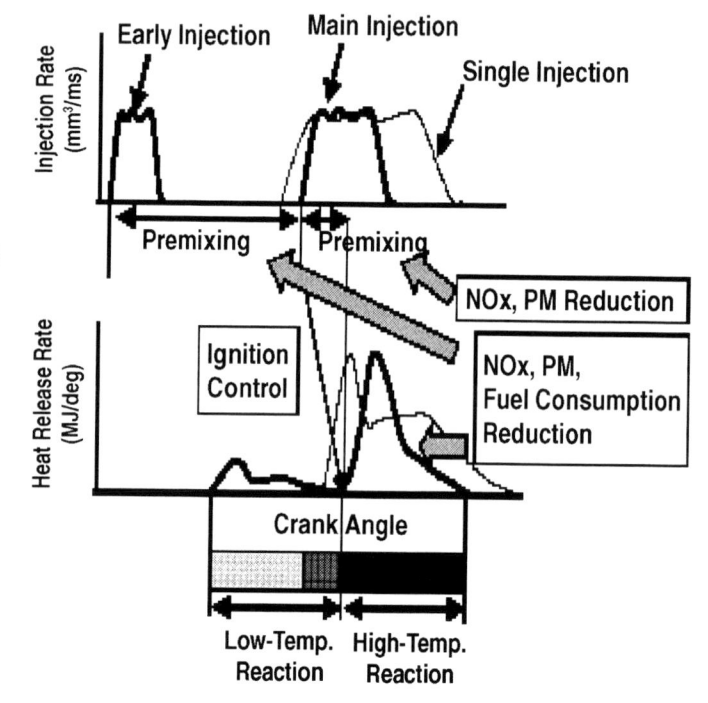

Looking beyond the two-pulse (split) injection, multiple injections of up to six pulses are now available, which may have additional potential for decreasing fuel consumption and emissions simultaneously. Figure 2.6 shows a proposed multiple injection strategy throughout the engine operation map, with three injections to attain a combination of low emissions, improved fuel economy, and power. At any particular engine operating point, the injection strategy can be optimized locally.

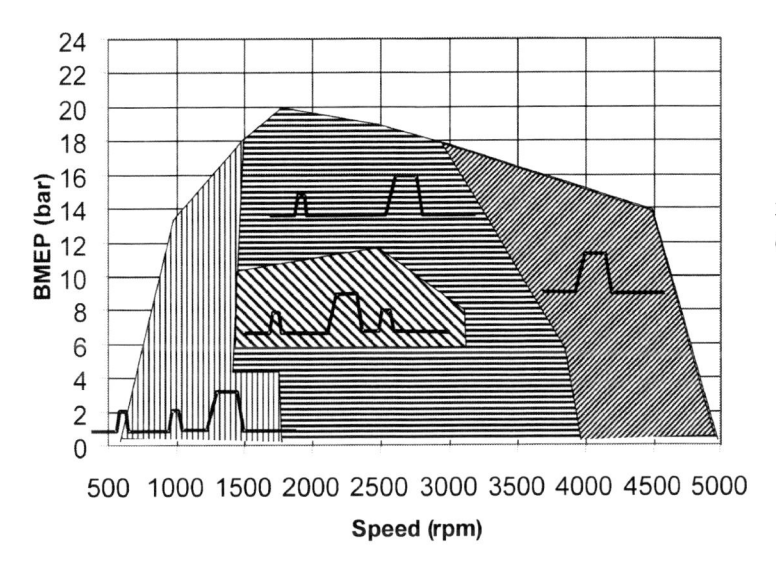

Figure 2.6 Example of a multiple injection strategy optimized over the engine operating map.

For example, using genetic algorithms in combination with 3-D combustion simulation, an optimal injection strategy of five injections for a high-speed diesel engine under a low-speed medium-load condition has been derived and is shown in Figure 2.7 [Ref. 2.3].

In addition to in-cylinder combustion and emissions control, the FIE may be further required to assist the exhaust aftertreatment by providing HCs for the regeneration of NOx and particulate filters or other catalyst device. The capability to electronically control their number, timing, and the post-injection fuel quantity independent of engine speed and load give the common rail system a key advantage over less flexible fuel systems. A proposed four- and five-pulse injection strategy for heavy-duty diesel engines with active aftertreatment regeneration is shown in Figure 2.8. However, there are problems of potential over-penetration of the fuel spray and consequent cylinder

Figure 2.7 An optimal injection strategy for five injections predicted using KIVA-3V code and genetic algorithms for a light-duty diesel engine at 1757 rev/min, 45% load condition [Ref. 2.3].

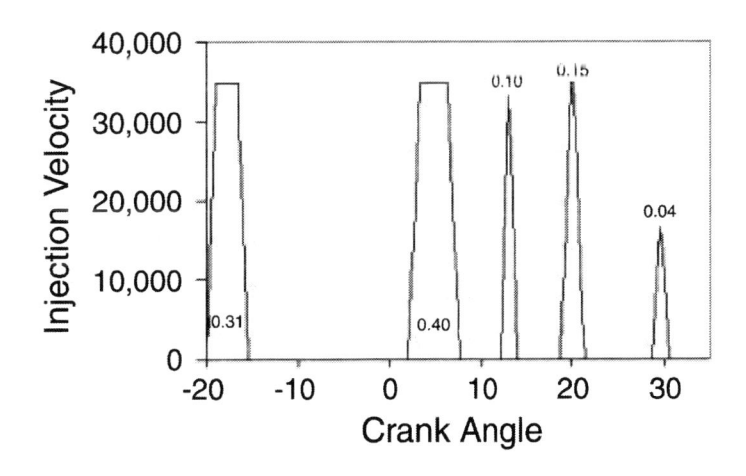

wall-wetting from post-injections late in the expansion stroke, because a rail pressure that is optimum for injecting into the dense air at TDC will now be too high for the lower density prevailing. Additionally, there is a risk of fuel contamination of the EGR system in cases where a high-pressure-loop EGR system is adopted.

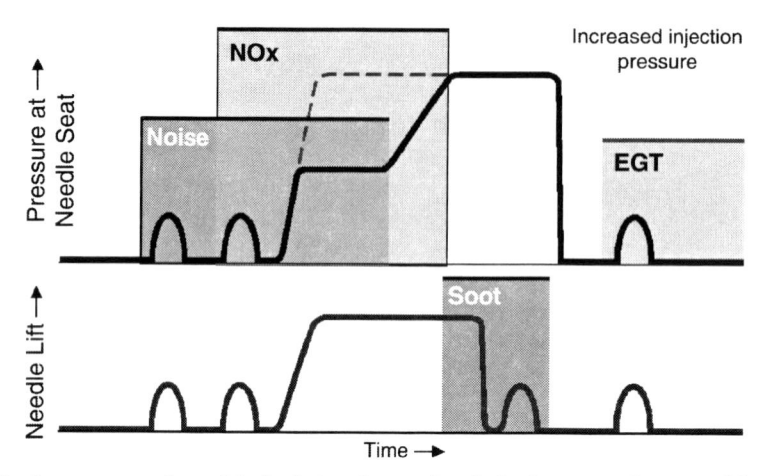

Figure 2.8 A proposed multiple injection schedule for an advanced heavy-duty engine, showing injection event regions that are helpful in the control of noise, NOx, and exhaust gas temperature (EGT) for aftertreatment regeneration.

The Nozzle

In the past, the long-stem hole-type nozzle for DI engines would be specified by the number of injection holes, their diameter, and the included spray angle, say, 6 by ⌀0.180 mm (diameter) by 150 degrees. Today, there is much more emphasis on nozzle-to-nozzle and hole-to-hole flow consistency due to their impact on exhaust emissions and performance. Thus, nozzles are now specified by number of injection holes and their included angle as in the past, but also by their flow value under standardized test conditions and the percentage hone that has been applied. In this test, the flow of calibrating oil through the nozzle at 100-bar pressure is recorded during manufacture, both before and after the abrasive hone preconditioning process, and this flow difference is expressed as the "percent hone." From this data, a coefficient of discharge (Cd) value can be derived, which may be as high as 0.9 (Figure 2.9). Because the hone operation is a time-consuming activity during nozzle manufacture, it is now common to produce tapered holes initially, which then require only 10% hone to achieve the required Cd value. Thus, a nozzle may now be specified, for example, as six holes, 150-degree included spray angle, and 1.45-liters/minute flow. It is, of course, also necessary to decide on the sac type: VCO or mini-sac. In the case of the latter, the sac diameter and volume may also be variables.

Nozzle hole with sharp edge entry
* Coefficient of discharge up to 0.7
* Pressure loss at hole entry
* Momentum loss in fuel jet

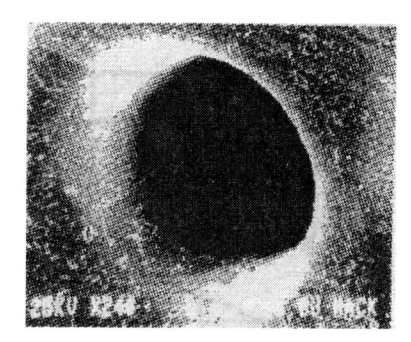

Nozzle hole with rounded edge entry
* Coefficient of discharge up to 0.9
* More momentum in fuel jet, thus better dispersion and atomization
* Made by hydro-erosion or extrude honing

Figure 2.9 Injection orifice entry conditioning. *(Courtesy of Delphi.)*

In common rail-type systems, the combination of injector needle dynamics and nozzle geometry determines the quantity of fuel metered and the final structure of the diesel spray. Nozzle configuration, with fuel injection pressure, ambient conditions, and fuel properties, are significant factors that affect spray characteristics, because the spray hole geometry influences the turbulence, cavitation, and velocity profiles of the internal fuel flow. The sac volume is the dead volume downstream of the needle valve inside the nozzle, which is exposed to the combustion chamber and may release fuel at the end of the combustion stroke, contributing detrimentally to the HC and PM emissions. Mini-sac or micro-sac nozzles (Figure 2.10) are designed to reduce the sac volume, but VCO nozzles have the smallest dead volume—about one-fifth that of the mini-sac nozzle. This is achieved by isolation of the sac in this type of nozzle, because the spray holes break through to the nozzle seat area, not into the sac.

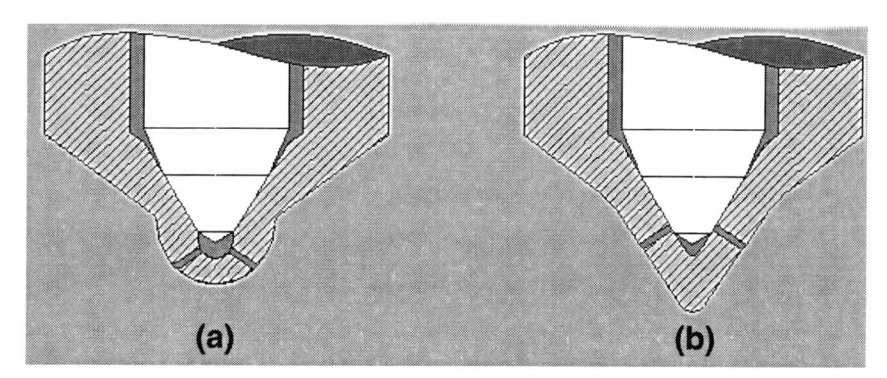

Figure 2.10 (a) Mini-sac and (b) VCO nozzle configurations.

Spray Structure and Breakup Mechanism

This section focuses on the dynamics of injection and spray, which are frequently used to gauge the performance of FIE. In general, the injection dynamics are controlled by the actuation of the needle and the preceding transient hydraulics from the rail to the nozzle tip; therefore, it is best understood by looking at the needle dynamics and the injection mass-flow rate. The resultant spray quality is controlled by the fluid mechanics within the nozzle tip and is best explained by spray visualization and other quantitative test bench measurements of spray symmetry, penetration, angle, drop size and velocity distribution, and so forth.

Breakup and atomization of liquid jets is a very complicated process. Many diverse jet atomization mechanisms have been proposed. Their mechanisms have been categorized by Reitz *et al*. [Ref. 2.4] as aerodynamic surface wave growth, pipe turbulence, rearrangement of the cross-sectional axial velocity profile of the jet, wall boundary layer exit velocity profile relaxation effect, cavitation, and liquid supply pressure oscillation. Simply put, flow instability in the liquid phase interacts with aerodynamic forces outside the nozzle to produce the diesel spray breakup process. Turbulence and cavitation have been identified as the main disturbances in the internal nozzle flow. Turbulence is generated by shear stress near the wall and at the hole inlet, and it has a strong effect on primary breakup and surface wake growth. However, the role of cavitation on atomization is less clear, but its transient behavior and interactions with exit velocity and pressure fluctuations certainly contribute to primary breakup.

Because of the fluid mechanics inside the nozzle, however, the VCO spray structure is quite different from that of the sac-type nozzles in terms of spray symmetry and plume included angle. In general, the spray from a VCO nozzle is puffier and less symmetric than its mini-sac counterpart, as shown in Figure 2.11. The mini-sac nozzle also can produce a more consistent and stable short pilot, although with poorer degraded atomization, as shown in Figure 2.12.

Figure 2.13 shows the difference in the internal flow structure entering the metering hole for these two classes of nozzle, with the velocity vector plots shown on the nozzle surface and symmetry plane, and on the plane of the hole axis. The sac volume functions like a buffer, producing a more even flow distribution among the spray holes. It also produces a smoother transition and more streamlined entrance into the metering holes. In contrast, the

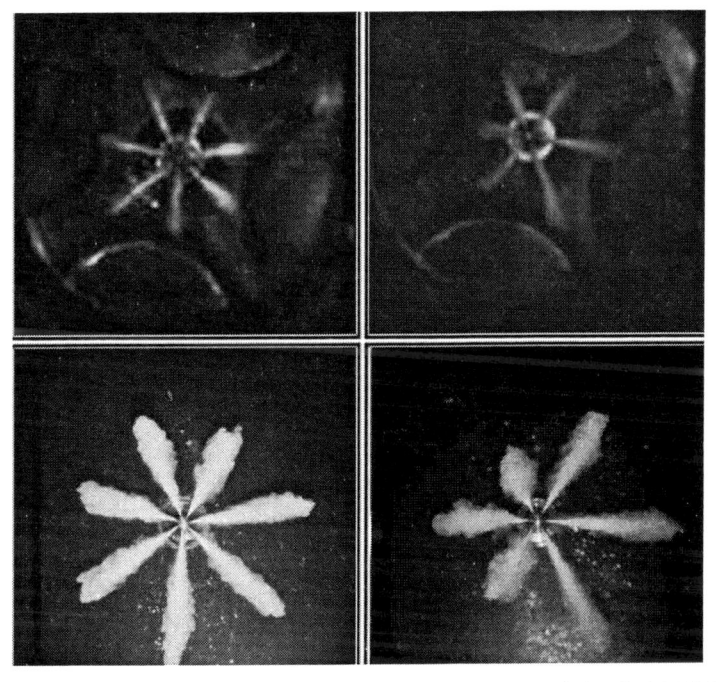

Figure 2.11 Comparison of sprays from (left) mini-sac and (right) VCO, taken under the same density conditions; (top) vaporizing (in an optical engine); and (bottom) nonvaporizing (in a pressurized chamber).

(a)

(b)

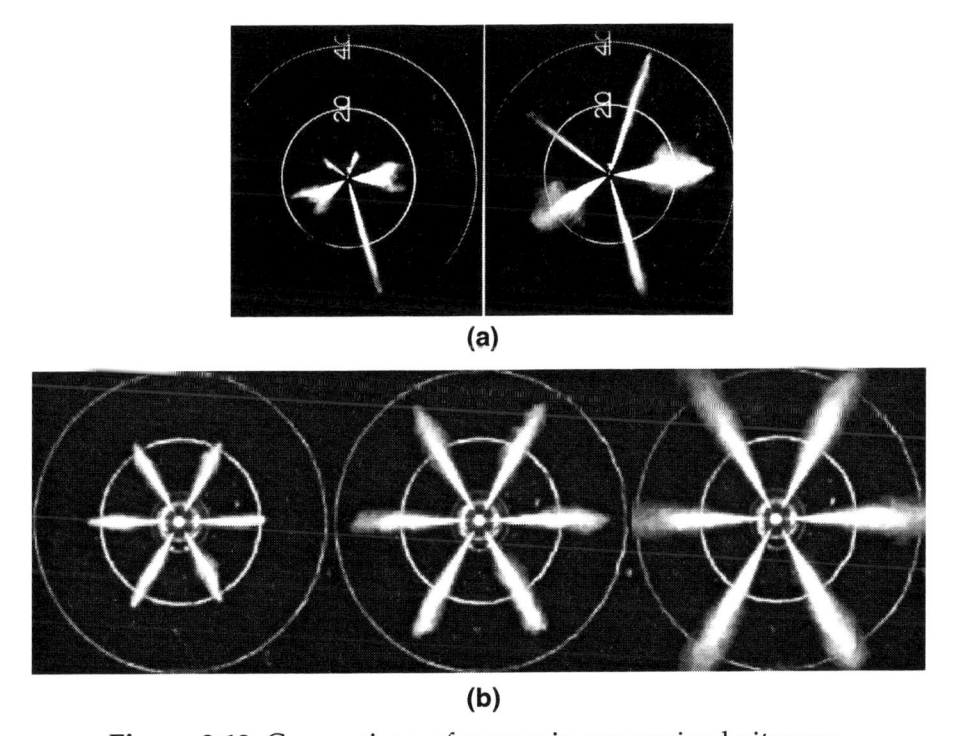

Figure 2.12 Comparison of sprays in pressurized nitrogen, with (a) bad and (b) good symmetry.

flow feeding the VCO nozzle hole is constricted between the needle and body seats, and it must make sharper turns and detours within that small clearance to achieve the transition. Therefore, the VCO nozzle generates more turbulence to assist spray atomization, although it is more susceptible to cavitation and flow separation and, in general, has a lower discharge coefficient than the mini-sac with the same hole diameter.

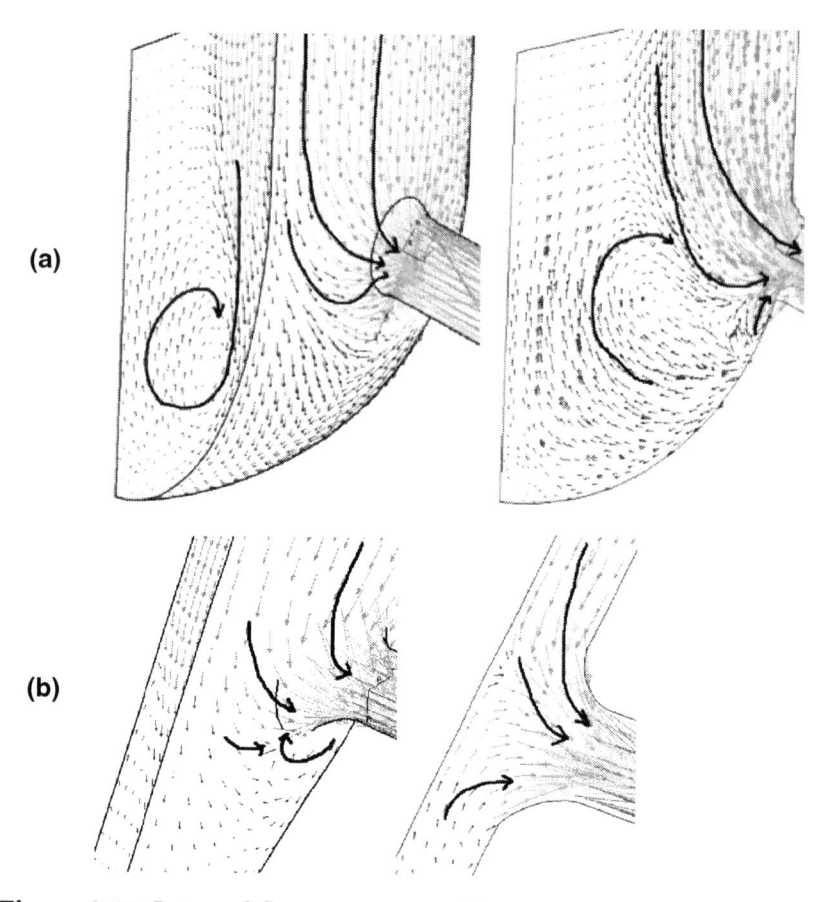

Figure 2.13 Internal flow structure of (a) the sac and (b) hole entry.

Cavitation is a very complicated phenomenon. Being the rupture of the liquid continuum due to excessive stress under highly dynamic conditions, it usually involves extreme pressure excursions. Statically, however, it occurs in a fluid when the local pressure drops below the vapor pressure, causing the fluid to flash-boil and create vapor bubbles or cavities. These then collapse when the pressure recovers, causing erosion damage to the component structure over time. Conditions that give rise to cavitation in diesel injector nozzles are both geometric and dynamic. Nonstationary characteristics of cavitation and pressure fluctuations will lead to exit velocity profile fluctuations. Soteriou *et al.* [Ref. 2.5] studied the effect of cavitation and hydraulic flip on atomization for actual-scale and scaled-up nozzles. Their results showed that (1) cavitation occurs in standard geometry diesel nozzles, (2) cavitation is the dominant mechanism of atomization, (3) cavitation results in the nozzle becoming choked, and (4) fuel subject to cavitation can be observed as an opaque foam exiting the nozzle. They also concluded that cavitation encouraged boundary layer separation, while increased turbulence at the nozzle upstream enhanced reattachment of hydraulic flip. The sharp turning angle inside the VCO nozzle is known to generate cavitation. Different cavitation patterns, including hydraulic flip, super cavitation, full cavitation, and cyclical cavitation, have all been identified. Figure 2.14 shows the simulated transient cycle of the growth and collapse of cavitation inside a VCO nozzle [Ref. 2.6].

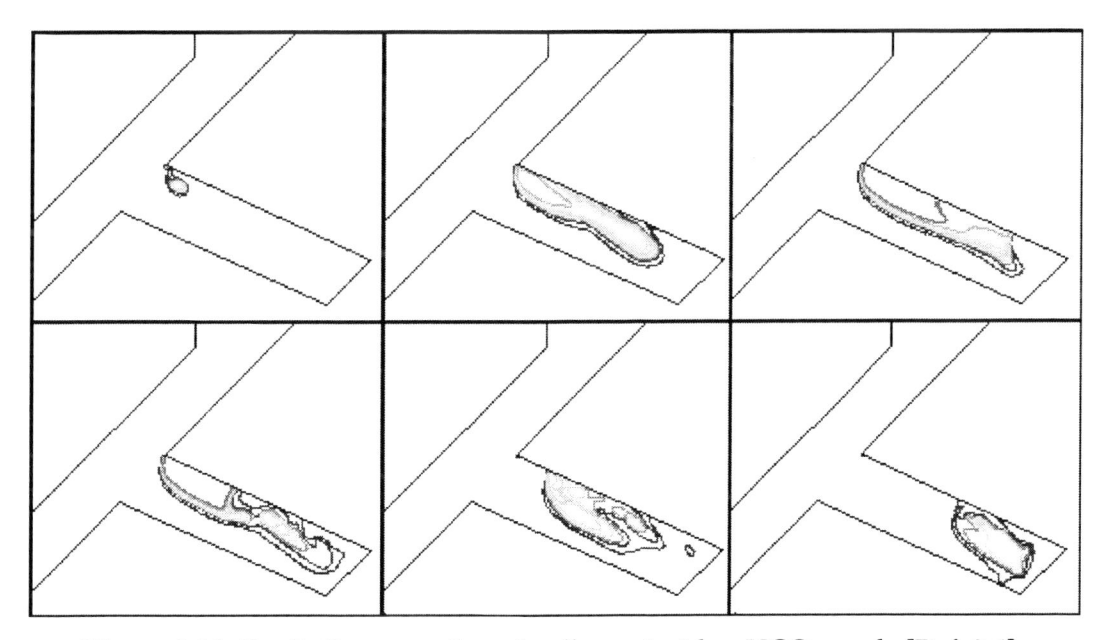

Figure 2.14 Cavitation growth and collapse inside a VCO nozzle [Ref. 2.6].

Because of the tight manufacturing tolerances necessary to meet the emissions warranty life of the engine, nozzles are usually preconditioned, so that the nozzle discharge coefficient remains stable over that period. This preconditioning is usually carried out through various abrasive machining processes, including hydro-erosive grinding (HG) and extrude honing, in which an abrasive-laden slurry is forced through the passages. The HG process is especially effective in conditioning the VCO nozzles by providing a smoother transition around the sharp corners and improving the entry conditions into the orifii. By minimizing or removing the vena contracta just downstream of the hole inlet, the net benefits are an increase of the mass flow rate of the effective hole area, reduction of flow tolerance, acceptance of higher injection pressures through lower material stresses, and prewear that provides greater performance stability over the emissions warranty period. Increasing the flow velocity in the injection nozzle spray hole is regarded as an effective way to obtain better fuel-air mixing without increasing the fuel injection pressure.

In combination with HG, tapered hole geometry (i.e., the entry is larger than the exit) is being used to improve jet flow characteristics (Figure 2.15). The basic principle is to provide a favorable pressure gradient inside the converging hole to keep out of the cavitation regime. Physical benefits are higher discharge coefficient (i.e., the same hydraulic flow quantity with a smaller geometric hole diameter) and reduced loss of pressure and fluid speed at the nozzle exit, particularly at low needle lift. In combination, these parameters have a strong influence on the primary breakup process of the liquid jet, as this manufacturing technology changes the properties of the flow inside the injection nozzles, particularly turbulence, cavitation, and velocity profile.

Needle concentricity in the area of the seat is a concern due to its effect on spray symmetry for VCO nozzles. In spite of the precision in manufacture, lateral articulation or "wander" of the needle tip is possible due to its distance from the guide diameter. To minimize this motion and the resulting preferential flow that causes asymmetric VCO spray plumes, dual-guided or extended-guide nozzles have been introduced. Figure 2.16 shows injector needle valve configurations with and without a dual guide. The repeatability or shot-to-shot variation of the injected fuel quantity, particularly at low deliveries, depends to some extent on the friction of the moving parts. As a

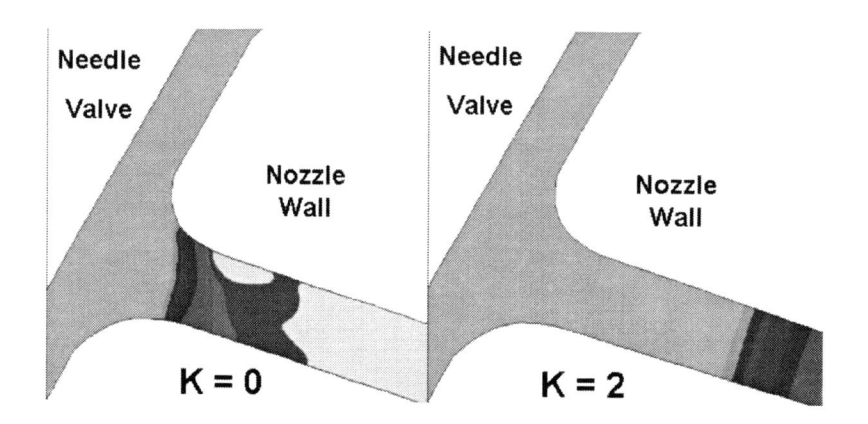

Figure 2.15 Computed pressure distributions inside VCO nozzle for a straight and a tapered hole. Tapering the nozzle hole has moved the location of the maximum pressure gradiant from the hole entrance to the hole exit.

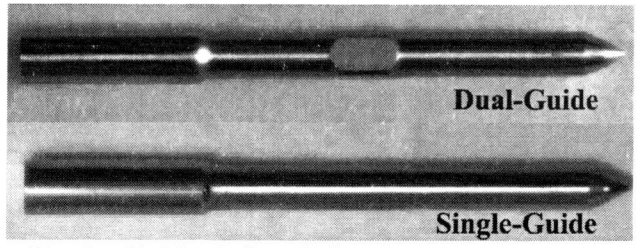

countermeasure, the nozzle needles may be coated with carbon or other low-friction material in the guide areas.

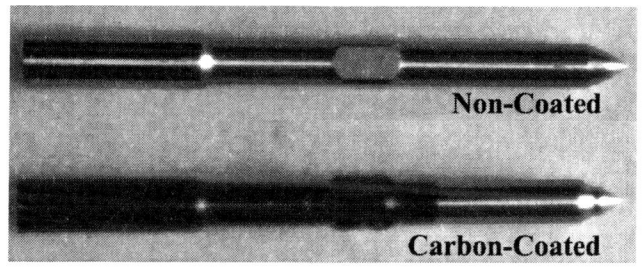

Figure 2.16 Injector nozzle needle valve configurations.

Figure 2.17 shows some recent nozzle valve tip seat geometries. For close tolerance of the pilot injection quantity, sophisticated needle seat designs, such as the grooved tip inverted (GTI, or ZHI in German) design, are currently produced. The upper conical seat area in the GTI design is confined by the cylindrical section above and the groove below, to provide maximum and minimum seat diameters. This effectively limits seat area migration due to wear over the nozzle lifetime. The lower conical seat area acts to reduce needle closing impact stress and wear on the upper seat, and to reduce dead volume. The effect of these design details is to reduce the delivery drift over time, and to improve the stability of tolerance on small fuel quantities for pilot injection. The figure makes a comparison between the standard double conical seat and the GTI seat. Figure 2.18 summarizes these recent nozzle concepts and features [Refs. 2.7 and 2.8].

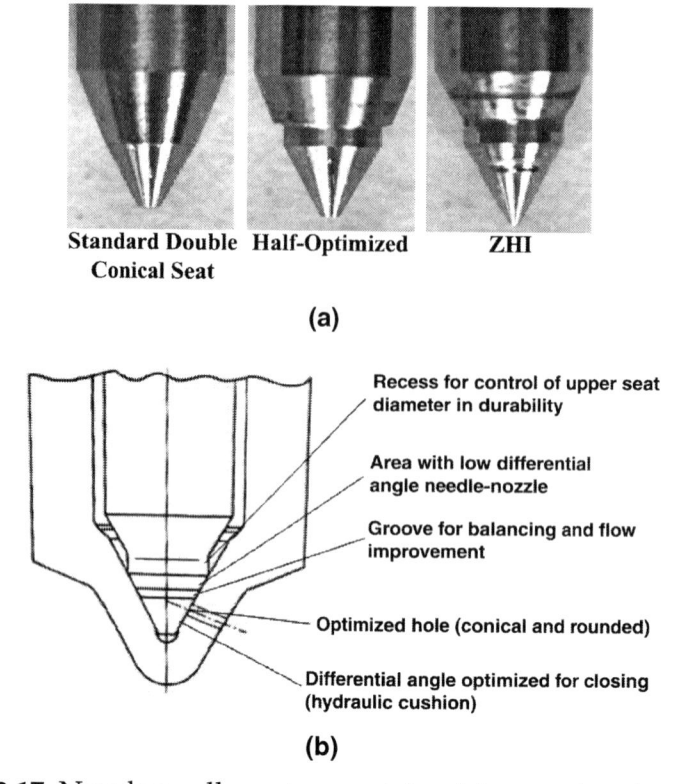

Figure 2.17 Nozzle needle seat geometries: (a) exemplary hardware, and (b) feature explanations.

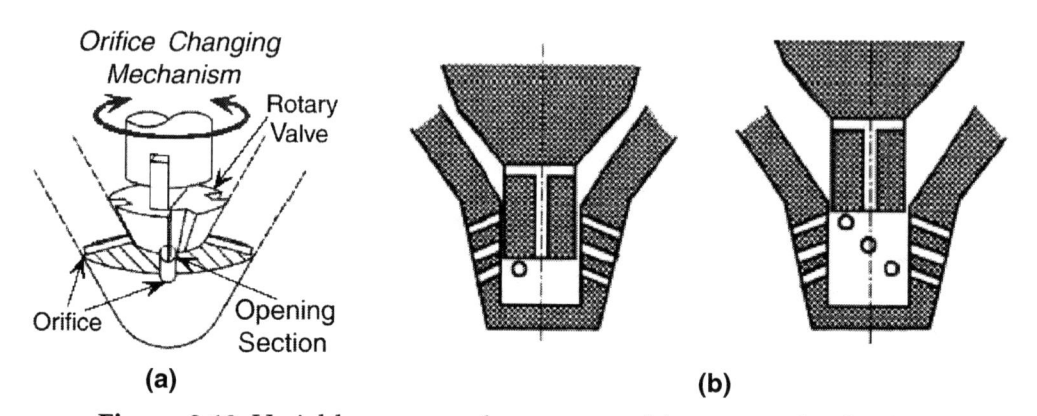

Figure 2.18 Variable area nozzle concepts: (a) rotary valve [Ref. 2.7], and (b) multistage [Ref. 2.8].

Variable Area Nozzles

Conventional nozzle flow area and geometry are chosen as a compromise between the specific requirements of power, consumption, emissions, and noise over the operating range of the engine. Within the constraint of peak pressure for which the injection system is rated, nozzle flow area is established from the required injection duration at peak power condition. At any speed/load condition below that, the flow area is larger than necessary, which has negative implications for spray quality. The additional freedom given by a variable nozzle area eases this constraint. When combined with an advanced injection system, it is possible to further optimize engine combustion and emissions performance, if the nozzle has the facility to vary the final orifice area at will across the speed and load range of the engine. Differentiation must be made between continuously (i.e., stepless) variable flow area nozzles (VAN), such as are available from some outwardly opening atomizers, and those variable orifice nozzles (VON) that vary it in, say, two discrete steps. Unlike two-stage injectors (TSI) in the past, which throttled the entrance of the nozzle holes to inject the fuel at a lower rate during the premix phase as a means to reduce combustion noise, VAN/VON should maintain a high spray velocity, thus minimizing smoke. All FIE manufacturers are developing nozzles in which the flow area and possibly even the spray targeting can be switched between at least two different conditions. Somewhat analogous to dual-length intake tracts, this will allow combustion to be optimized at light-load conditions, or perhaps under homogeneous charge compression ignition (HCCI) operation, and still deliver rated performance when called on to do so. A number of different geometries are proposed to achieve the required effects, such as rotating the needle to open different sized holes (Figure 2.18(a)) and multiple-stage nozzle tip (Figure 2.18(b)), and single tip with double needle designs, which open different numbers and diameters of complete holes at different times.

The Bosch coaxial vario nozzle, shown in Figure 2.19, uses a piezoactuator to control two coaxial intermeshing nozzle needles to open two superimposed rows of spray holes. The first row of holes with fine orifices for low flow is deployed at light load and may be optimized in terms of hole number and spray cone angle to reduce combustion noise in general and to reduce emissions. When approaching full load, the second row of spray holes, possibly with a different orifice diameter, is opened as well.

Figure 2.19 Bosch vario nozzle tip.
(Courtesy of Robert Bosch.)

Chapter Three

Common Rail Fuel Injection Equipment

The term "common rail" has, in common parlance, become debased and, in a general sense, has come to suggest an advanced technology diesel fuel system that may or may not follow the traditional common rail fuel injection equipment (FIE) architecture. Thus, the term has become somewhat generic in nature and does not differentiate among the various architectures that are sometimes ascribed as such. In other words, there is a strict definition for common rail, and then there is a loose definition.

The strict and traditional definition describes a system in which there is a high-pressure pump that raises the fuel pressure in a single stage, from the nominal feed pressure up to the specified system injection pressure. This fuel is then distributed to the individual injectors via the convenient disposition of a distribution gallery or rail. Injection timing and quantity metering are effected at the injector by mechanical (traditional) or electronic control means. This type is sometimes referred to as high-pressure common rail (HPCR), but this can still be misconstrued because all systems are high pressure in some respect (Figure 3.1) [Ref. 3.1].

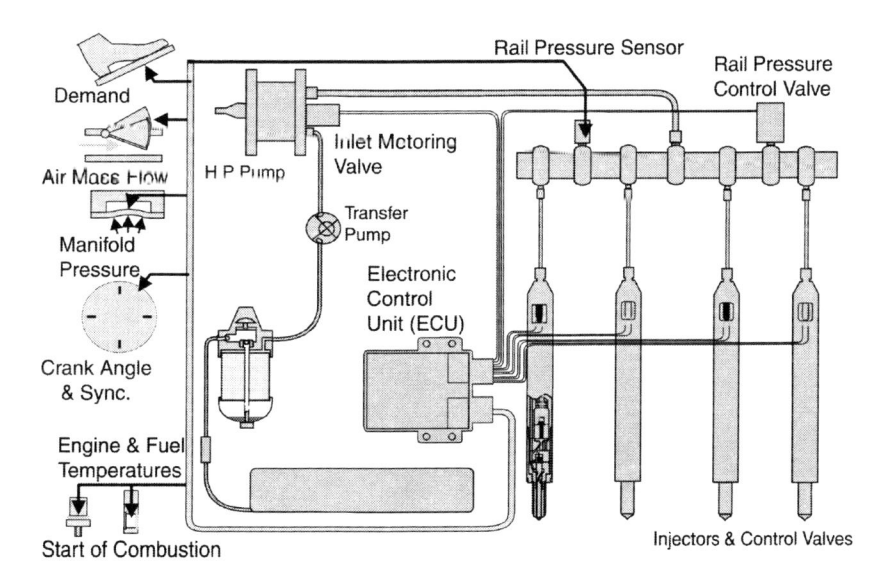

Figure 3.1 Schematic of a high-pressure common rail (HPCR) system [Ref. 3.1].

Also frequently embraced by the common rail term are the intensified systems (such as the hydraulic electronic unit injector or HEUI, described in Chapter Four), because they have a common pressurized gallery to which the final injection pressure is responsive. However, there are sufficient differences in both system architecture and functionality that, except under the most broad-brush description, it is misleading to refer to them as common rail systems. Similarly, there have been occasions in the general media where mechanically actuated electronic unit injectors (EUI) have been misconstrued as common rail on the basis that it is an advanced FIE technology with a common low-pressure fuel feed gallery to each injector. The problem arises because in a technology-savvy market, the end consumer is interested in what makes the latest product superior to the previous generation. However, that same consumer cannot be expected to understand all the arcane features that differentiate the competing FIE systems. Thus, in advertising and media reports, it is easier to lump them all under the common rail banner, which has thus come to represent advanced diesel engine technology.

Nevertheless, one cannot afford to be too pedantic about generic use of a name. As fuel systems evolve in the future, they may converge on some form of intensified common rail (ICR), in which case, the term "common rail" will still be generically correct.

Rarely, if ever in the past, have diesel FIE concepts impinged on the consumer's consciousness as they have today. In part, this is due to the fact that the modern reincarnation of the common rail concept has brought a step-change to the capability of diesel FIE and therefore, in turn, to diesel engine performance. This has been most apparent in the case of light-duty (LD) automotive engine performance, because it has enabled the wholesale shift from indirect injection (IDI) combustion systems to direct injection (DI) predominance. Prior to the advent of common rail FIE, available fuel systems were reasonably well matched to the requirements of the prevailing divided-chamber combustion systems, but only marginally commercial when applied to light-duty DI engines. When compared with the market standards established by the gasoline engine, deficiencies were apparent in terms of specific power, fun-to-drive factor, idle noise, and other refinement parameters such as exhaust odor and smoke. Some highly developed rotary pumps were able to show the potential of DI combustion systems but were expensive to manufacture. Also, because of their pump-line-nozzle (PLN) architecture, they still were not able to deliver the refinement necessary to make the diesel engine, for the consumer, transparent versus the gasoline engine.

Although common rail as an FIE concept today appears to be an obvious technology path, it begs the question: "Why did the major FIE companies not develop it sooner?" The answer lies in part in the fact that high-speed actuator technology necessary for control of multiple injections was not readily available until the late 1980s. Such actuators had been developed and productionized for EUI, but they were too large and expensive for light-duty use. A secondary reason is that as history has shown, development of new FIE systems is a long and expensive proposition and not to be undertaken lightly. All FIE manufacturers had huge manufacturing investments in PLN and EUI technologies, and much of that capacity would be redundant in the face of a major switch to the new technology. This consideration counseled a cautious approach to radical technology. Perhaps then, it is not surprising that modern attempts at common rail were initiated by vehicle OEMs having no in-house FIE manufacturing background and thus no existing capacity to lose.

One such little-known initiative was undertaken by General Motors in the United States in the early 1980s at the time that Oldsmobile was developing its V8 and V6 divided-chamber light-duty passenger car diesel engines. However, the origin of the modern HPCR system concept that is so prevalent in the European light-duty sector is due to the R&D groundwork performed at Fiat Central Research (CRF) in the early 1990s, and by Ganser in Switzerland. However, FIE suppliers Denso pioneered modern common rail FIE systems for medium-duty engines, and L'Orange for heavy industrial-sector engines.

Market Drivers

Fundamental factors behind the development of common rail FIE can be identified in the European governmental desire to minimize traffic congestion, environmental pollution, and foreign oil imports, all of which were becoming problematic in the 1970s as the postwar economy grew. These desires resulted in attempts to modify public behavior by progressively raising motor fuel taxation, which in turn caused consumers to seek transport solutions that offered ever-better fuel economy without giving up mobility performance and sociability. Initial attempts at practical diesel cars during the 1930s through the 1960s were marginal in their functional effectiveness, but by 1970, diesel cars were a practical proposition with IDI combustion systems, particularly in France and Germany. Comparison of similarly specified gasoline and diesel-engined cars generally indicated a 15 to 20% fuel economy advantage to the latter, with little meaningful penalty in performance

However, the same societal pressures also made the DI diesel look like logical technology path to follow because it offered a further 10 or 15% economy improvement. The problem was that no suitable injection system existed for this class of engine. All light-duty fuel systems had been developed primarily for divided-chamber combustion systems, for which peak injection pressures of approximately 350 bar seemed adequate. Early attempts to use truck-engine-derived FIE for light-duty DI engines typically lacked the peak pressure capability necessary for competitive power density, or alternatively, could not deliver the refinement expected by this market sector. Nevertheless, this market demand encouraged the development of several new light- or medium-duty EUI fuel systems, such as the Bosch injector developed for Volkswagen, the Lucas product for the Land-Rover TD-5 engine, and the Caterpillar HEUI adopted by International Truck & Engine for the Ford "Powerstroke" engine. Concurrently, a new generation of high-pressure rotary-type pumps was developed by Bosch and Lucas, but they were the last and expensive gasp of the long PLN heritage. Truly, common rail FIE has been a game-changing technology.

The higher-pressure capability of common rail FIE relative to previous rotary and in-line FIE, and the rate-of-injection flexibility provided by injection pressure control and multiple-injection capability, have made the current ascendancy of DI diesel engines possible. Additionally, this performance has been delivered at a cost that is much lower than could have been contemplated in the early 1990s with precommon rail technology, even had it been possible. There is also the comfort factor that common rail system architecture is similar in concept to the well-known and understood gasoline injection system, which therefore makes it easier to accept. Further, being modular in nature, there is an expectation (probably more imagined than real) that an engine OEM will be able to build a low-cost system by "cherry-picking" the cheapest modules from the various FIE suppliers.

High-Pressure Common Rail

We refer here to the systems in which the rail pressure is essentially the same as the injection pressure. This is the dominant concept currently with systems manufactured by Bosch, Delphi, Denso, and Siemens, in frame sizes that span the range from light-duty automotive, through medium-duty on- and off-highway, up to heavy-duty on-highway. Variations on this same theme are also available for large industrial and medium-speed marine engines. Clearly, the benefits provided by common rail fuel systems are being recognized across the board by all sectors of the diesel community.

This recognition has brought a sea-change in the FIE industry, as all available resources that normally would be working on the steady development of conventional FIE have now been drafted into product and manufacturing engineering of this new product. Within a very short space of time, the conventional PLN FIE was being superseded by the new technology and relegated to a "back-burner" status. In practical terms, this has meant that common rail manufacture now occupies the flagship plants that often are located conveniently close to the product engineering or headquarters locations, and the previous mainstream products have been moved to lower-cost Third-World factories. In and of itself, this is not a new phenomenon because as the global standards of manufacturing gradually improve, what once were cutting-edge products that needed careful engineering "nurse-maiding" will, at some point, approach commodity status that less exalted factories can essay. This same evolution will occur with common rail systems, too.

Here, we should revisit the basic HPCR system architecture, before examining the constituent modules in more detail. Fuel is drawn from the tank, filtered, and delivered to the high-pressure pump. The output flow, and therefore pressure from the pump, is dynamically modulated to align with the engine demand, and this fuel then charges the rail. Injectors fed by jumper pipes from the rail incorporate actuators, which control the injection timing and the delivered quantity, all under the management of an engine control unit (ECU) and the usual array of sensors (Figure 3.2).

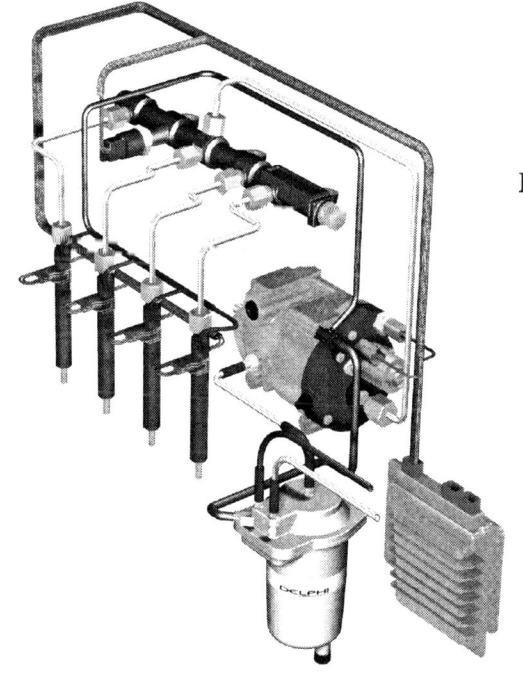

Figure 3.2 An automotive mechanization of a common rail system.
(Courtesy of Delphi.)

Low-Pressure System

Easily overlooked, but only at one's peril, the low-pressure system embraces the fuel tank, filter, on occasion the feed pump, pipe work up to the high-pressure pump, and any return flow piping. It is not unusual for more irritating warranty and service-related problems to be due to low-pressure system issues than the much more expensive high-pressure system.

Whereas it is almost universal for gasoline fuel systems to utilize in-tank electric fuel feed pumps, it is less common for diesel fuel systems. A concern for safety is the usual reason given, because under worst-case conditions, the fuel vapors in the tank can fall within the flammable range, whereas in the gasoline tank, they are always too rich. Under normal conditions, the low volatility of diesel fuel results in the vapors being too lean for ignition, but this can change under adverse conditions, particularly in the event of misfueling. Fuel feed pump options therefore are a carefully insulated in-tank pump, a conventional external pump, or an engine-driven mechanical pump, frequently available as an integrated option within the common rail high-pressure pump. In the case of external pumps, note the need to minimize the number of joints in the pipe work between the tank and pump because the system is under a vacuum and will cause service problems should it lose its prime. Further, the pump should be designed to remain flooded after shutdown, so that in the event of partial overnight drain-back, the pump will immediately prime on initial crank. This is easily ensured by arranging for the inlet and outlet connections to the pump to be on the top, not underneath.

Fuel waxing is an issue for diesel vehicles during winter months. The temperature at which the wax precipitates out of the fuel to the point where it clogs the feed line is very dependent on the specific fuel properties. As a result, there are different fuel grades for winter and summer months. They may also vary from region to region as a function of anticipated weather conditions. A summer grade fuel may wax at −4°C, whereas winter grades are blended to be wax-free to perhaps −12°C or lower. Because the mobility of vehicles means that any vehicle could be operated in a subzero climate, it is normal practice to install an electric fuel heater in the most troublesome locations in the fuel system. In virtually all cases, this is at the inlet to the fuel filter because wax crystals will not pass through the paper element, thus causing a fuel blockage. Nevertheless, small-bore pipes with sharp bends or other wax-trapping features should also be avoided on the suction side. Fuel waxing in the tank is rarely a problem, because once the engine is running, warm fuel recirculated from the high-pressure pump and injectors is returned to the tank, or if a temperature-responsive diverter valve is specified, it can be returned directly to the filter.

As peak injection pressures escalate with each new generation of fuel system, so must the precision of manufacture of the wetted components exposed to that pressure, if parasitic leakage is not to assume unacceptable proportions. Extremely small diametral clearances, precision valve lifts, and extended durability expectations driven in part by long emissions warranty periods all make the systems less tolerant of dirt and demand ever tighter standards of fuel cleanliness during service. For these reasons, the fuel filtration requirements for the common rail class of FIE are more demanding than any previous generation. Additionally, most wetted FIE components are necessarily made from steel and therefore must be protected from water or other corrosive fluids, such as the aqueous urea used by the selective catalytic reduction (SCR) system. The fuel conditioning package therefore may comprise a fuel heater, a fuel filter, a differential pressure sensor, a water separator and trap, a water level sensor, and a temperature-responsive hot return fuel diverter valve. Fuel heaters typically are manufactured from a positive-temperature-coefficient (PTC) ceramic material, having self-regulating current characteristics, so that no additional thermostat and relay are required. A typical fuel conditioning package is shown in Figure 3.3.

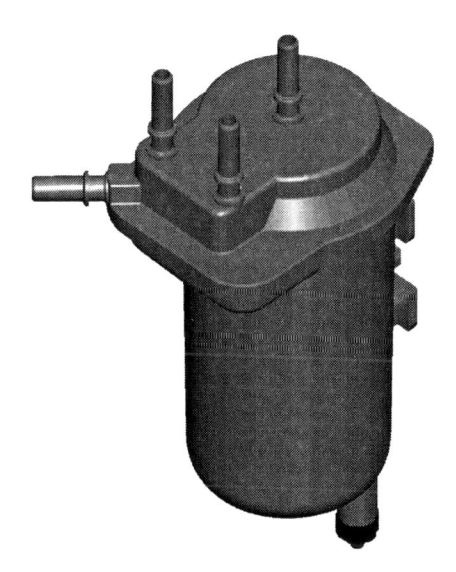

Figure 3.3 Typical fuel filter canister.
(Courtesy of Delphi.)

The sensitivity of modern fuel systems to dirt is such that they are assembled in clean-room conditions at the manufacturing plants, packaged carefully for transport, and require similar levels of cleanliness during the engine assembly process. For engine OEMs, this can be an onerous requirement if cleanliness has not been an observed practice; however, lack of care taken at this stage in the plant is repaid with early life and warranty failures that will cost the company dearly. The fuel filter and everything downstream from it must be assembled with this extreme level of attention to cleanliness, including the pipe from filter to high-pressure pump that is usually installed at the engine assembly plant.

Fuel filtration standards have been encoded in ISO standards, where it can be seen that those that apply to common rail fuel systems are more severe than those for earlier-generation FIE (Figure 3.4). Other important fuel filter requirements address filter service change intervals, end-of-life disposal, and filter behavior under vehicle crash situations because the filter is usually mounted accessibly in the engine compartment.

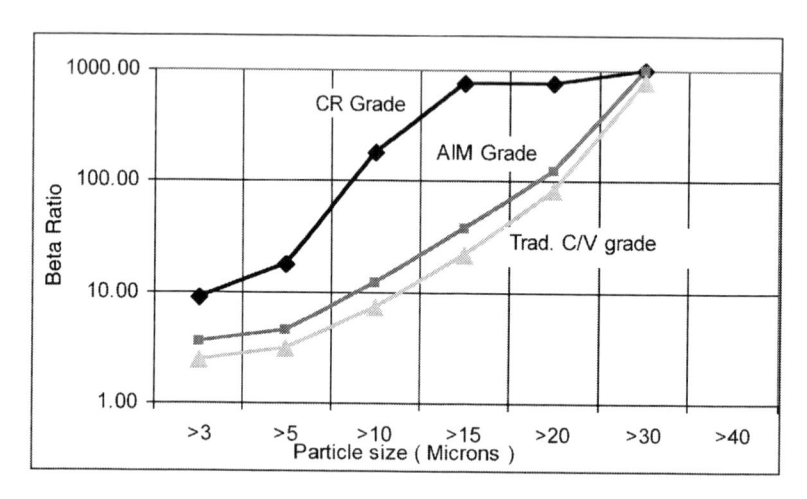

Figure 3.4 Typical ISO/TR 13353 filtration curves.
CR = common rail; AIM = industrial; C/V = commercial vehicle.
(Beta ratio = particle count [of given size] before versus after filter.)

High-Pressure Pump

A characteristic of the injection system under consideration in this chapter is that the rail pressure is elevated in a single stage, that is, there is no area-ratio intensification involved between feed pressure and injection pressure, simply a piston pump. The work involved in raising the fuel to be injected from feed pressure (say, 3 bar) up to injection pressure (say, 1600 bar) is significant and is on the order of 4 kW or more for a 2-liter engine at 4500 rev/min. To contain these pressures and absorb that amount of power means that the pump will necessarily be a bulky, robust unit. Additionally, the cyclical nature of the pumping process is likely to result in a drive torque signature with high positive and negative amplitudes that act as a prime noise generator; thus, care must be exercised in any pump-mounting arrangement.

Excluding certain industrial engines that are likely to utilize one or more single-plunger pumps, the bulk of the automotive engines are likely to select either a radial (star) or an in-line configuration for the high-pressure pump. As in previous eras, light-duty applications have selected rotary radial-piston pumps, as exemplified by the three-cylinder Bosch/Rexroth-type star geometry pump (Figure 3.5(a)) or Delphi internal cam-ring pump (Figure 3.5(b)). Heavy-duty applications often choose in-line pumps, as typified by the Cummins "CAPS" two-cylinder pump. The typical star geometry pump is a single-bank, equally spaced two- or three-plunger design that is reasonably compact, having a pancake-like form-factor. A single eccentric on the driveshaft strokes the plungers through a slipper mechanism that relies on the fuel properties for lubrication. Engine-oil lubrication may be an option for certain application or geographic variants. A through-the-pump driveshaft affords the option of ganging an additional device, most likely a transfer pump, onto the back face.

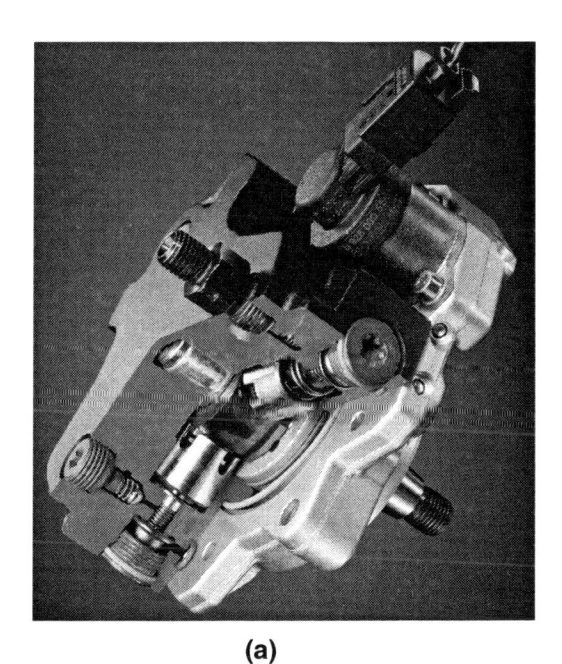

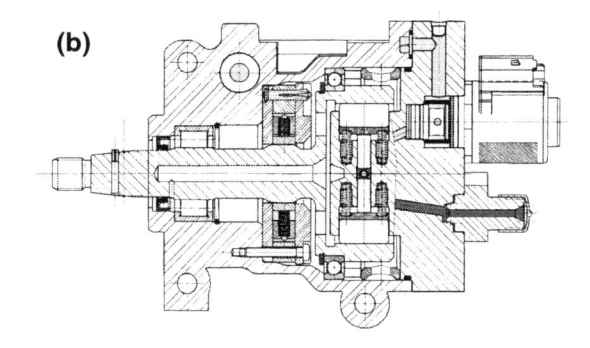

Figure 3.5 (a) Three-cylinder, star-format Bosch CP3 high-pressure pump (end view). (b) Delphi internal cam-ring opposed plunger high-pressure pump, with integrated transfer pump (side view) [Ref. 3.1].

The pump, as a self-contained module, can incorporate several subsystems within its envelope. The prime function obviously is to generate a fuel flow quantity as derived from pedal-demand, and fuel pressure as requested by the calibration map. However, beyond that, it is common for the pump also to include an integrated transfer pump, which may take the place of the upstream feed pump. That is, the transfer pump draws fuel under suction from the tank, thus obviating the upstream feed pump, unless a particularly high head is involved. Also, the high-pressure pump is likely to incorporate a flow-metering valve that is under ECU control, which meters fuel into the high-pressure stage in proportion to the engine torque and injection pressure demands [Ref. 3.2]. Other less complex features may include pressure regulators for both transfer pressure and over-pressure safety release.

Arrangements for driving the high-pressure pump may vary from one engine design to the next, and this variation must be catered for by the FIE manufacturers. Thus, it is typical to find accommodation for belt, chain, gear, or direct drive. In the case of belt drive particularly, substantial bearings must be packaged within the pump to accept the high side loading applied to the driveshaft. For direct-drive applications, the couplings may be of the polygon, spline, or Oldham coupling variety. Drive speed ratio between the pump and engine is typically, but by no means always, 0.5:1 for light-duty automotive applications. However, one of the key deciding factors is the desire to synchronize where possible the pumping pressure pulses to the injection events for optimum line-to-line fuel quantity balance. This can pose problems in the case of a three-cylinder pump when applied to, say, a five-cylinder engine. While an approximate phase relationship between pump and crankshaft may be specified as suggested above, the accuracy is greatly relaxed in comparison with a conventional injection pump installation.

Pump drive torque signature is a key parameter that can have major repercussions on the drive selection and design. A smooth signature will extend drive belt life, or minimize gear rattle, as the case may be. There are also implications on the pump mounting arrangement, because the pump can be a noise, vibration, and harshness (NVH) exciter unless rigidly mounted to a robust damping structure. A flywheel mass is often attached to the pump driveshaft to minimize the torsional perturbations imparted into the drive system. Many new diesel engine designs have chosen rear accessory gear trains, where the drive originates from a crank nodal point, having the dual benefit of a near homokinetic input signature to the pump, and the robust mounting suggested above. A three-bolt ISO standard flange is the most common mounting.

Pumps will normally be available in a family of displacements to suit the range of engine applications. At minimum, the pump must displace the maximum fuel demand from the engine, plus system leakage with assumptions for high-mileage degradation; however, a relationship between pump displacement and system volume also exists. To meet time-to-first-fire targets, it is necessary that few crank revolutions are required to raise system pressure to the minimum operational level for starting. Additionally, instantaneous fuel flow demands must be met during load transients. Because of the high levels of energy required to pressurize fuel in modern injection systems, all future pumps will feature a variable displacement capability so that no more energy is absorbed than is absolutely necessary. In most cases, this will be achieved through the medium of an electromagnetically controlled inlet-metering valve that restricts pump filling and thus volumetric displacement. In this manner, under all part-load conditions, the pumping plungers do not travel their full stroke. Given sufficient sensitivity of control, as can be obtained through a PID feedback loop, pump output in terms of delivery and thus pressure can be contrived to very closely match the engine demand. Under these circumstances, pressure will normally have to be dumped only from the rail under a down-transient condition. This represents a major savings in energy and a reduction in heat rejection, relative to these first-generation HPCR systems that featured fixed displacement pumps. Rail pressure may be controlled between a minimum value of perhaps 250 bar at low idle up to the rated value that could be invoked at any high-load condition above, say, 1000 rev/min, particularly if the calibration calls for high exhaust gas recirculation (EGR) flow rates. However, the deployment of high injection pressure at low speeds and loads should, as a general rule, be resisted if possible, because the high pumping parasitic under those conditions will result in poor brake specific fuel consumption (BSFC).

Rated system pressures across the industry seem to have fallen on a common set of incremental steps, which almost fall along design generational lines. Original systems were released at 1350 bar, followed by 1600-bar systems, with 1800- and 2000-bar systems on the near-term technology road maps (Figure 3.6) [Ref. 3.3].

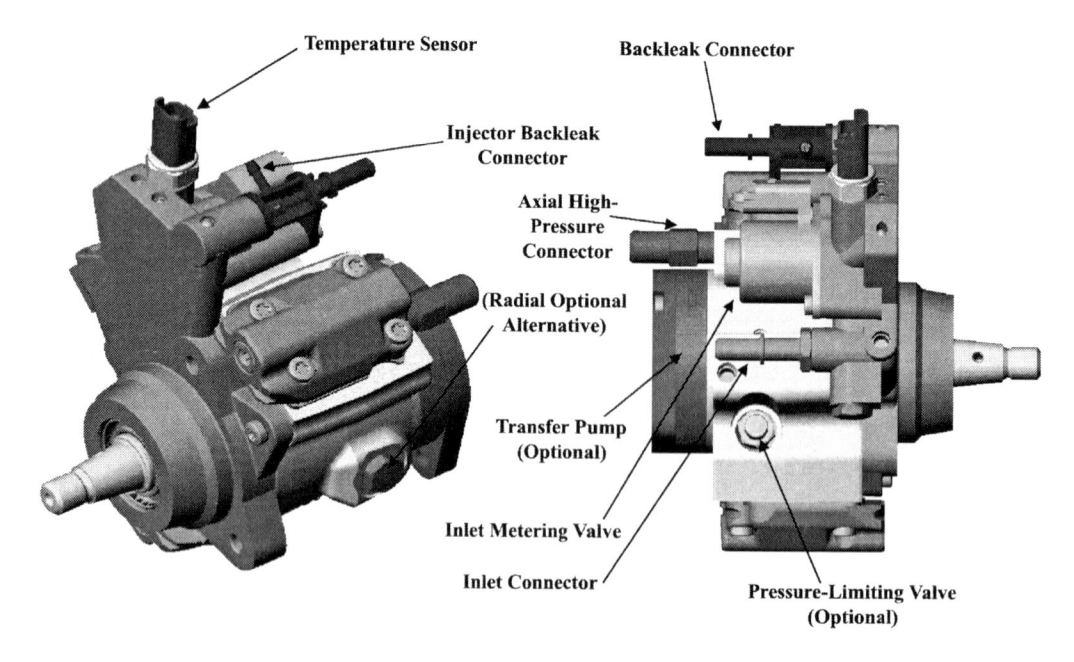

Figure 3.6 Delphi 2000-bar-capable high-pressure pump [Ref. 3.3].

The rated system pressure has significant implications on the high-pressure pump design and specification. Higher pressure ratings will result in higher drive torque requirements and higher contact loadings among plunger foot, tappet or slider, and cam follower. Seat loadings for pumping-chamber valves will be higher, as will separation forces on sealing joints. For the typical light-duty application having a fuel-lubricated pump, it is the cam/cam follower/plunger interface that presents the most serious tribological issues with the low-sulfur fuels of the future, and this is a key driver for industry-desired fuel lubricity standards. The U.S. Environmental Protection Agency (EPA) mandated low-sulfur fuel having a maximum level of $S = 15$ ppm will become effective as of June 2006, and because this will be achieved in many cases through additional hydrotreating, the lubricity can be expected to be marginal. The FIE manufacturers have, as of this writing (2005), adopted a common position on fuel lubricity, calling for a high-frequency reciprocating rig (HFRR) acceptance wear-scar value of less than 0.460 mm on European fuels. Higher values than this are indicative of fuels having unacceptable lubricity to deliver the required fuel system durability in service. Nevertheless, in North America, ASTM has now included a lubricity specification in the ASTM D 975 standard for diesel fuel at the 0.520-mm HFRR wear-scar diameter level, which is a reasonable near-term compromise between the FIE manufacturers' desires and those of the petroleum industry. This became effective on January 1, 2005.

Because the high-pressure pump absorbs considerable power, the heat that is generated in the translation of rotary motion into pressure energy must be dissipated; thus, all such pumps have a through-flow of fuel for thermal management. This flow is tapped from the transfer pump output and is directed to the cam chamber, where it both lubricates the internals and picks up excess heat, before venting back to the filter in cold weather or to the tank in hot weather. Heat buildup in the fuel tank can be an issue under worst-case conditions of high ambient temperatures, high power demands, and low in-tank fuel levels. For some applications, this can obviate the use of plastic fuel tanks or require a fuel heat exchanger.

The Rail

Having generated the fuel pressure, it is, of course, necessary to distribute it in a safe and convenient manner, with a sufficient reserve so that transient flow demands on the system can be met. The traditional approach to this problem has been to provide a tubular manifold that runs the length of the engine in similar fashion to the fuel rail of a conventional port-injected gasoline engine; however, the different order of pressure—2000 bar versus 3 bar—demands a totally different product.

The diesel fuel rail is typically a forged steel product with integral connectors for fuel inlet from the pump and multiple outlets to the injectors. Multibank engines will have a separate rail per bank with a common distribution block in the center. Tubular, or "log" design rails are the norm, but so-called spherical rails with the ports radiating in an octopus fashion also are available, with selection being based principally on packaging issues (Figure 3.7). Internal fuel volume is determined by the ability to smooth out the pressure excursions excited by pump output and injector demand pulsations, both of which benefit from a large rail volume, and, on the other hand, rapid rail filling and pressurization at cranking for quick starting that benefits from a smaller volume. Rail volume therefore is an application variable, and a volume of approximately 10 cc's per liter of engine displacement is not uncommon.

Figure 3.7 Examples of spherical and tubular fuel rails.
(Courtesy of Delphi.)

Because rail pressure is a key control parameter, it is logical for the rail pressure transducer to be located on the rail in such a position that it records a mean value that is not unduly affected by internal pressure waves. It is usual for the transducer to incorporate self-contained electronic signal conditioning that provides an analog voltage output.

It has been pointed out that under "up" transient conditions, the rate of rail pressure increase can be important; however, a similar but opposite situation can arise during a "down" transient. In this case, rail pressure may be higher than is desired for the light-load condition that occurs on throttle tip-in following a deceleration, and in this case, rail pressure must be bled off. For this purpose, some systems have a pressure regulator valve on the rail to dump pressure under ECU control when necessary, and to provide a fine control on pressure under normal running conditions. Common rail injectors with the more agile balanced control valves are able to dispense with this regulating valve on the rail. Through multiple rapid short-duration pulses of the injector control valve between intended injection events, it is possible to drain off rail pressure into the return line without invoking an unintended injection.

High-Pressure Pipes

The high-pressure pipes are a very important part of any diesel injection system, but particularly so in the case of common rail systems, because they are expected to be entirely problem free in a high-vibration environment while being continuously exposed to high internal pressures. No leakage can be tolerated, because even a pinhole can produce a skin-piercing spray, with the serious prospect of blood-poisoning for the unwitting mechanic.

Too easy to overlook beside the other sophisticated system components, the high-pressure pipes are a key entry point for dirt anytime they must be removed, and it is common practice to renew the pipes at this time. Pressure seal is normally made on conical 60-degree metal-to-metal nipple seats retained with M14 nuts (Figure 3.8). It is important to keep all rail-to-injector pipe lengths identical on an engine, but pump-to-rail pipe length is also important because it can sensitize the system to pressure wave effects, although snubber orifii at the rail inlet and outlets are sometimes employed.

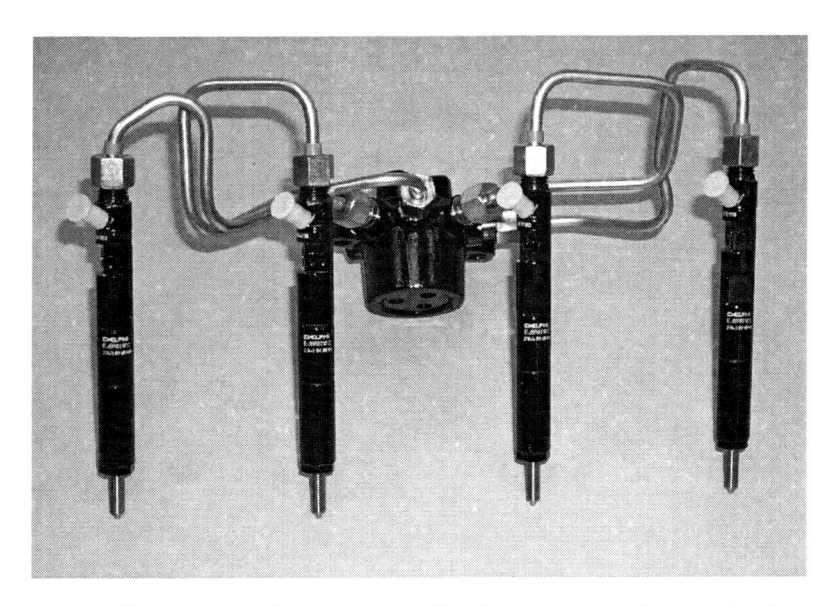

Figure 3.8 Rail, pipes, and common rail injectors for a four-cylinder engine.
(Courtesy of Delphi.)

Common Rail Injector

The common rail injector is the key component of the system, and the performance of this device largely differentiates one manufacturer's system from another. The more flexible and responsive the system and principally the injector is, the easier the calibration engineer can bend it to his or her will in delivering a world-class engine with customer-pleasing drivability and sociability. The injector conventionally is located in the cylinder head and integrates the injection control valve and the atomizing nozzle. In the HPCR concept, it is usual (although not universal) for the rail pressure to be continuously present at the nozzle, and only precision manufacture of the nozzle assembly prevents unintended leakage of fuel past the nozzle needle seat and into the combustion chamber. Any such leakage would adversely affect hydrocarbon (HC) emissions or, if excessive, could cause serious engine damage; hence, the need for extreme cleanliness and fine fuel filtration will be appreciated.

Parameters that can be quantified on test bench or on engine that provide the product differentiation include shot-to-shot consistency, minimum stable pilot fuel delivery, minimum separation between multiple injections, injection-to-injection interaction, stability and linearity of delivery gain curves, peak injection pressure capability, and rate of injection control, particularly on nozzle closing. All of these aspects are determined essentially by the injector (Figure 3.9).

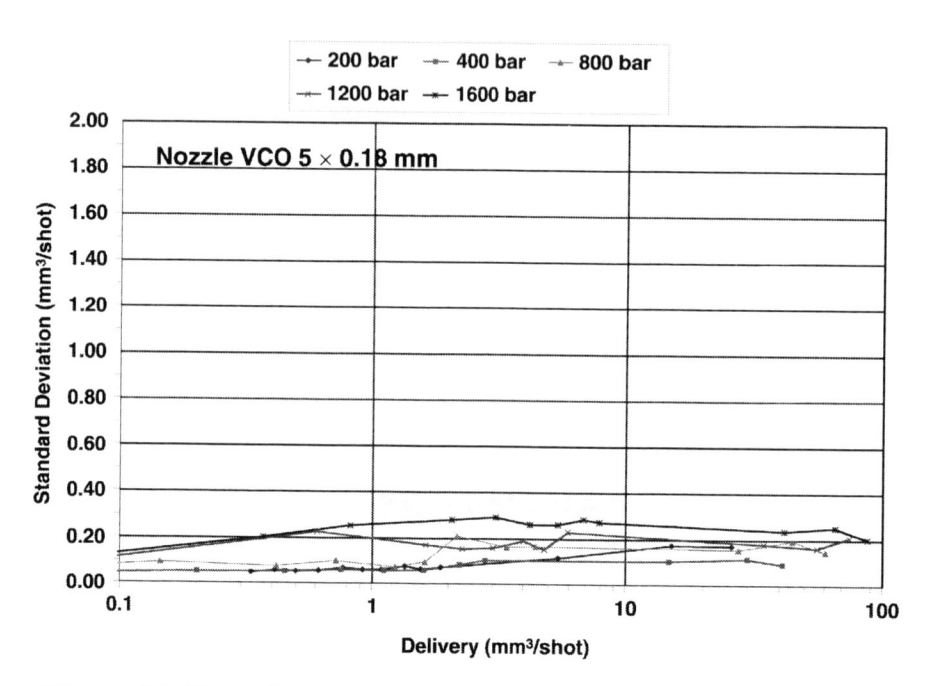

Figure 3.9 Typical standard deviation curves for injected quantity.

Also of importance is the rate of fuel backleakage from injector to tank, because this directly correlates to wasted energy and thus fuel consumption. Indeed, this return flow of fuel can present a heat-rejection problem, sometimes requiring the addition of an expensive fuel cooler, as recited previously.

As also mentioned previously, an additional key differentiating parameter is the injector form-factor, which can impose restrictions either on the cylinder head design, if the injector is selected first, or on the choice of injector, if the cylinder head design is fixed. Generally, the most difficult application to accommodate is the small DI engine with the now-conventional four perpendicular valves with central vertical injector, and a cylinder size below about 0.3 liter. In this situation, the room remaining for the injector is extremely limited, so that specification of anything but the slimmest of injectors is likely to lead to performance or durability compromise. Even with larger cylinder sizes, a slim injector, if able to deliver the desired durability and functionality, can provide more room for larger inlet ports or cooling passages.

With the common rail injector now retained in the cylinder head, it is necessary to make connections for high-pressure fuel, low-pressure fuel return, and the electrical signal. Examining the high-pressure fuel connection first, there are generally two options: top-mounted axial inlet, or some arrangement of side inlet feed. In turn, either option can involve a conventional 60° cone, nipple, and pipe nut, or a clamped connection. In the case of clamped side-feed designs, a conical depression in the side of the injector is loaded by a perpendicular intermediate tube extending through the side of the cylinder head. Fuel from the rail thus is conducted to the injector through this tube, and a separate clamp arrangement holds the injector into its pocket. Although not currently done, it is possible with axial inlet injectors to combine both injector retention and the fuel connection with a single clamp, as was done cost effectively in the past on precommon rail Saurer engines.

At some point in the future, it is possible that many engines may have "dry overheads"—that is, either no valve gear lubrication system or one that is independent of the dirty crankcase oil [Ref. 3.4]. Until that time, however, it is necessary to guard against the possibility of lube oil dilution resulting from fuel leaks at the injector or its high-pressure connection. Sometimes, this is done by arranging for the high-pressure connection to be external to the valve cover; sometimes, it is done by isolating the injector pocket so that any unintended leakage is captured with the regular leak-off in a drain drilling that eventually returns it to the tank. In some cases, the return fuel flow from the injectors may be captured in metal pipe work under the valve cover, or alternatively with plastic pipe work if external.

With respect to the electrical connection, there will be variations among different vendors' systems, depending on the injector design details and the actuator technology. In general terms, solenoid actuators require low voltage but high current levels, and piezoelectric actuators require high voltage but lower currents. This has some influence on the type of connector used.

Actuator location on the injector is another obvious visible differentiator. First-generation common rail injectors, with the singular exception of Delphi, all mounted the actuator on top of the body, remote from the nozzle (Figures 3.10 and 3.11). In large part, this was due to the selection of a non-balanced control valve that required high operational force and therefore a large and powerful actuating device too big to submerge within the injector body. On smaller engines, the actuator therefore would protrude above the valve cover where the ancillary connections could be made, but which was an unwelcome addition to total engine height. More important, however, this construction necessitated a very long control piston in the form of an internal spindle extending the length of the injector body from the control valve at the top down to the nozzle at the bottom. The role of this spindle was to transmit the biasing force of the servo piston to the nozzle needle. However, because this spindle of some considerable mass is obliged to follow the desired motion of the needle, it must of necessity detract from the responsiveness of the injector. This dynamic limitation appeared for engine calibrators as a constraint in

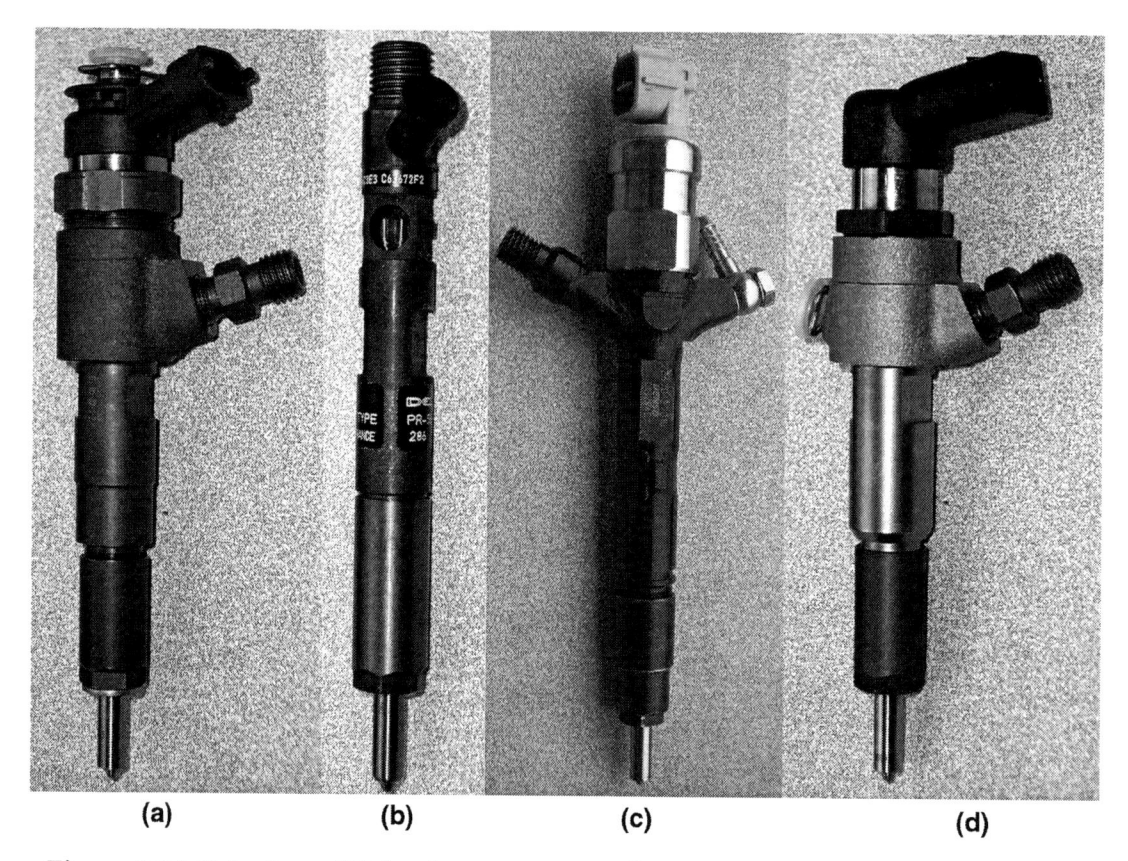

Figure 3.10 Selection of light-duty common rail injectors of different form-factors: (a) Bosch, (b) Delphi, (c) Denso, and (d) Siemens.

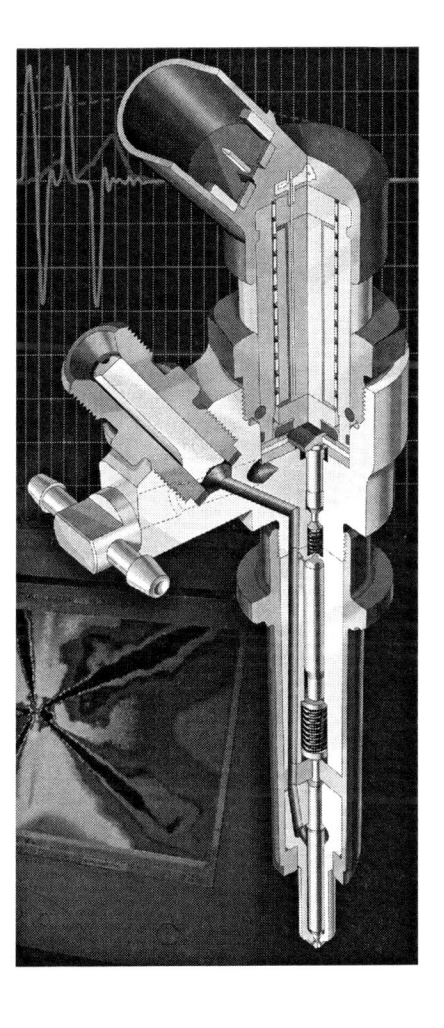

Figure 3.11 Siemens piezo common rail injector section, with control valve detail.

achieving the desired small pilot injection quantities and minimum pilot-to-main separation times. Therefore, there is a strong incentive to follow the Delphi lead and bring the control actuator into as close proximity as possible to the nozzle. In so doing, a reduction of dynamic mass is effected, and a valuable improvement in response is obtained.

In general, this path has been taken by all FIE manufacturers in their second- or third-generation products, but it is essentially a packaging challenge because the control valve and actuator must fit within the constraints of the Ø17- or Ø19-mm body. Allowance also must be made for the fuel to travel the length of the body, because this construction allows for a top-feed axial inlet injector design, and this will normally mean that the actuator is offset from the center of the body if the fuel drilling passes down one side. There is, however, also a worthwhile reduction of radiated noise, because the powerful actuator mounted external to the valve cover is now replaced by a lower-force unit submerged within the cylinder head.

Control Valves

A key parameter of future diesel fuel injection systems will be, as it has in the past, the level of control that the system affords to the engine calibrator. The term "control" requires definition, but precision with respect to time, speed, and repeatability are all important aspects of valve performance that have improved with each new generation of product. Generally, there are two technical solutions to translating actuator motion into nozzle needle motion. Servo action is by far the most common, but direct actuation has been shown to be possible in conjunction with piezoelectric actuators [Ref. 3.5]. Early attempts at direct actuation with solenoid actuators as exemplified by the Sofredi system were not commercially successful.

Where high fluid pressures are involved, as they are in any modern injection system, leakage of that pressure is to be avoided or, if not possible, then certainly minimized. Once energy has been expended to generate pressure, then any loss, however small, has a deleterious effect on the total energy balance and thus the vehicle fuel consumption. Usually, the location having the highest parasitic leakage in the system is at the injection control valve, and there are several different approaches to this problem. Additionally, it must be recognized that injection pressure in the nozzle sac is a more important parameter than rail pressure, because the former takes account of any dynamic flow losses that occur to the fuel on its passage through the injector. Moreover, injection quantity metering in common rail systems is a function of the fuel pressure at the nozzle, the flow area at the orifii, and the time duration over which that pressure is exposed to the orifii. Thus, for accurate fuel metering, stability of system pressure is desired, along with consistent actuator performance.

Two-Way Valves

A fundamental choice to be made is that between two-way and three-way servo valves, and current systems employ both types. In the two-way valve, there are two ports, one of which is connected to the control pressure and the other to the system drain. For the relatively long duration between injections, the control valve is closed; therefore, there is no leakage flow from this source. However, when injection is desired, the valve is opened, allowing the control pressure to dump to drain pressure; thus, there is a continuous loss of fluid for the duration of the injection event. The volume of fluid lost is a function of the prevailing pressure in the system at that time and the size and flow characteristics of the control orifice upstream of the valve. A balance exists between the size of the control orifice and the response rate of the injector. Generally, a larger orifice flow area will provide a faster response, but the trade-off is higher leakage flow and thus greater parasitic loss. Certain designs aim to minimize this parasitic leakage by arranging for a pintle on the control piston to close the drain port at full valve lift. A discussion on the pros and cons of two- and three-way control valves in diesel FIE application may be found in Ref. 3.6.

Three-Way Valves

Turning now to the three-way valve, this design has three ports: one for the supply pressure, one connected to the controlled entity, and one for drain. The fundamental advantage of the three-way valve over the two-way valve is that there is no continuous loss of fluid during the injection event because the servo control port is connected to the pressure or the drain port as appropriate, with the other port being isolated. In most cases, the pressure port is connected to the control port for the long duration between injections, and it is then isolated during the injection event while the control volume is allowed to drain through the other two ports. On the face of this, it would appear that the three-way valve is the most appropriate for the task at hand because it should deliver the highest efficiency. However, in the relatively simple valves that, due to ease of manufacture, are generally found today, the transition losses can be high and sometimes higher than that of a well-designed two-way valve. It will be appreciated that these transition losses occur when the valve moves from one position to the other. In so doing, there is a short period of time when all three ports are in common communication; therefore, system pressure has direct access to drain. Parasitic pressure loss thus is dependent on the valve design in question and the speed of the associated actuator. Three-way valve designs are available that will close the pressure port before opening the drain port, and vice versa, but such valves come at the expense of manufacturing complexity and cost.

Balanced and Unbalanced Valves

A further design feature that can be applied to either of the foregoing servo-valve types is the choice of balanced or unbalanced valve, and again both types are found in current injection systems. Widely used because of its simplicity, the unbalanced valve operates in a force-balance mode, in which a spring typically holds the valve onto its seat, in opposition to the system pressure. It is necessary that an actuator be chosen having characteristics that can provide the necessary response when required to pull in against this spring force. A reasonable compromise can be found between actuator bulk, force requirements, electrical drive requirements, response time, and hydraulic system pressure by the careful selection of control valve seat diameter. Typically, the control volume bleed orifice size determines the rate of injector needle opening. In turn, the highest contemplated system hydraulic pressure when applied to this orifice area determines the minimum valve seating load required. Simple as it may be, this configuration does involve significant compromise, and one effect is that control valve actuation response times may vary as a function of the prevailing system pressure. In turn, this may complicate the engine mapping and calibration process if it results in fuel delivery versus drive pulse duration gain curves that are not relatively linear, as illustrated in Figure 3.12.

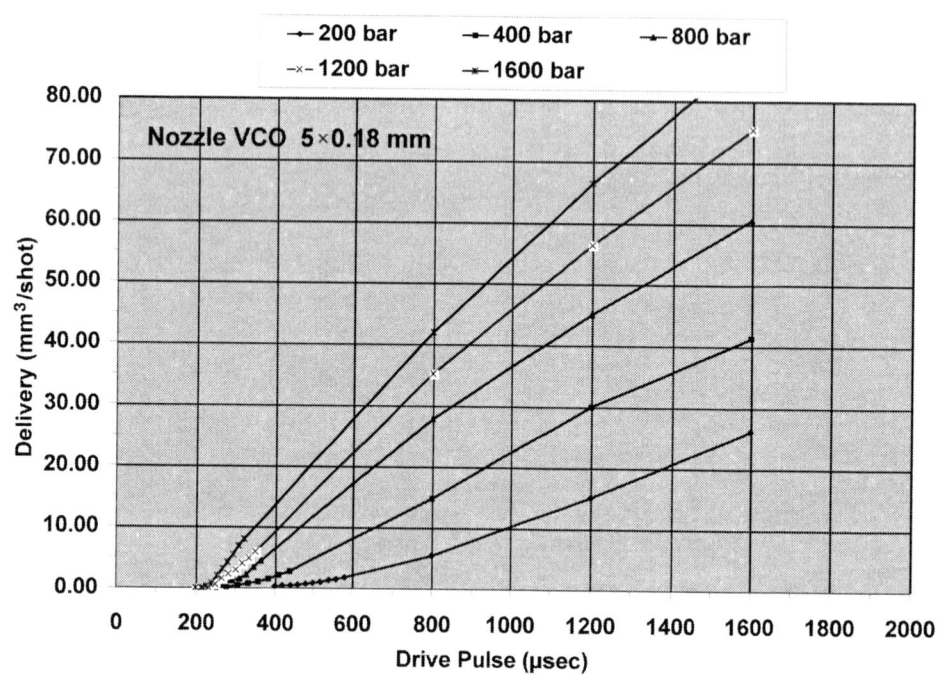

Figure 3.12. Typical "gain" curves for fuel delivery versus drive pulse duration.

An inherent characteristic of the unbalanced valve is that in the event of a system failure that results in an over-pressure condition, it can be arranged for the hydraulic force as applied to the control valve to overcome the opposing spring force. This will result in opening of the valve and the safe bleeding-off of pressure, before the onset of system structural failure (Figure 3.13).

As implied by its name, the geometry of the balanced valve is so arranged that when pressurized, the hydraulic forces are applied uniformly to the spool between the piston and valve seat, so that there is no resultant axial force component (Figure 3.14). This means, therefore, that the force required

Figure 3.13 Unbalanced two-way control valve operation: (1) control valve on seat, (2) control valve off seat, and (3) control valve returned.

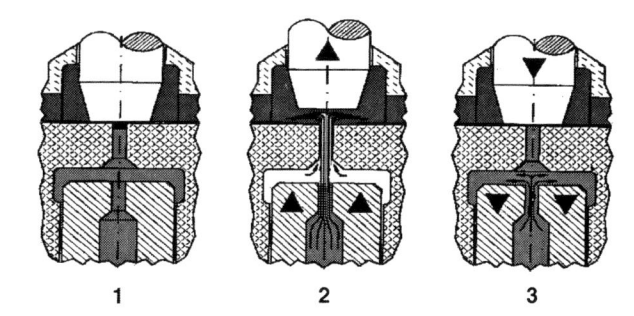

1. Not injecting: Solenoid not energized, valve closed typically by a light spring.
2. Solenoid energized: Needle valve opening rate controlled by orifice size.
3. Rapid closure: Solenoid de-energized, control piston repressurized.

Control valve closed / Nozzle closed

Control valve open / Nozzle closed / Fuel discharge via CV

Control valve open / Nozzle open / Fuel injected

Control valve closed / Nozzle closing

Control valve closed / Nozzle closed

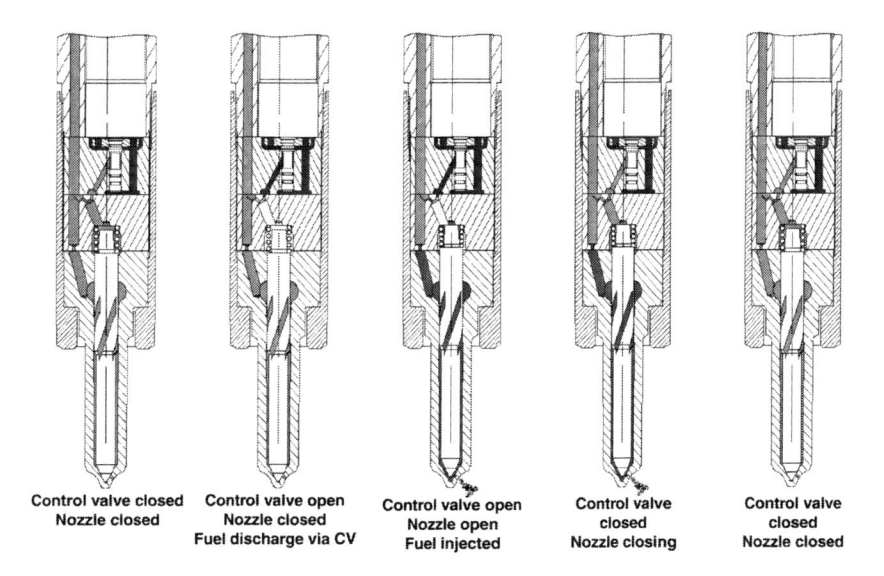

Figure 3.14 Delphi balanced control valve actuation sequence.
(Courtesy of Delphi.)

to move the valve is typically much less than is the case for the unbalanced valve. Thus, the actuator parameters such as size and electrical requirements to achieve the same response times may be smaller. Again, these distinct advantages result in an elegant solution but may come at the expense of slightly greater manufacturing complexity and therefore of initial cost. However, system cost may be lower, because the reduced electrical current demand of the balanced valve permits lower-cost drive electronics.

Digital latching control valves, as typified by the Sturman design (Figure 3.15), require balanced valves as a fundamental feature to achieve their bipolar behavior. Residual magnetism is employed to hold the valve in one state or the other, and in this way, electrical energy consumption is markedly reduced in comparison to conventional peak-and-hold waveforms, because no hold-on portion is required.

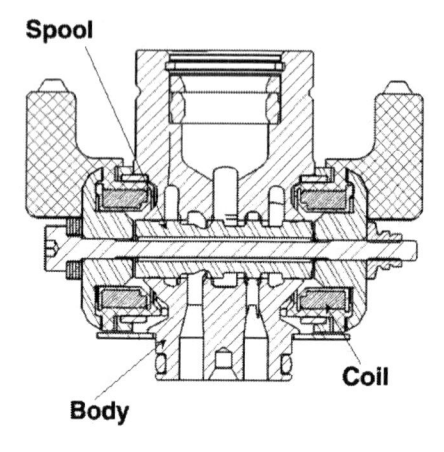

Figure 3.15 Sturman digital solenoid control valve.
(Courtesy of Sturman Industries Inc.)

The design uses a double-acting solenoid architecture, in which the spool serves double duty as both the hydraulic control valve and the solenoid armature. Typically configured as a hollow-spool valve with several porting configurations available, the valve has high dynamic response due to the low mass of the spool, plus the absence of a return spring [Ref. 3.7]. Moreover, valve response can be modulated on demand as may be desired to achieve injection initial rate control, by selective timing of energization of the opposing solenoids. Magnetic forces, however, can be relatively low with this construction, resulting in susceptibility to shot-to-shot variation due to dirt or friction. Some degree of spool position feedback information can be extracted from the inductive signal provided by the inactive coil, which may be used for control purposes.

To date, this technology has been released in conjunction with HEUI-type FIE where it is used to admit oil at rail pressure (~250 bar) to the injector intensifier. In this application, it has demonstrated itself to be a compact and elegant solution.

Actuator Technology

Given that diesel engine performance is so dependent on the fuel quantity control and injection timing characteristics of the fuel system, it will be recognized that the injector control-valve actuator is a key technology. Currently, the two actuator technologies in use for FIE are electromagnetic solenoid and piezoelectric ceramic devices, the latter being significantly more expensive (Figure 3.16). Solenoids predominate at present, but piezoelectric is fully expected to supplant solenoids for many applications by 2007 and eventually perhaps most applications as the escalating emissions standards change the cost-to-benefit ratio. For best-in-class performance, a design solution is sought that combines the quick response of a low-inertia piezoelectric actuator in conjunction with a fast-response servo mechanism, or better still, a direct connection to the nozzle needle. In turn, the system response depends on the total inertia of the actuator, servo mechanism, and nozzle needle.

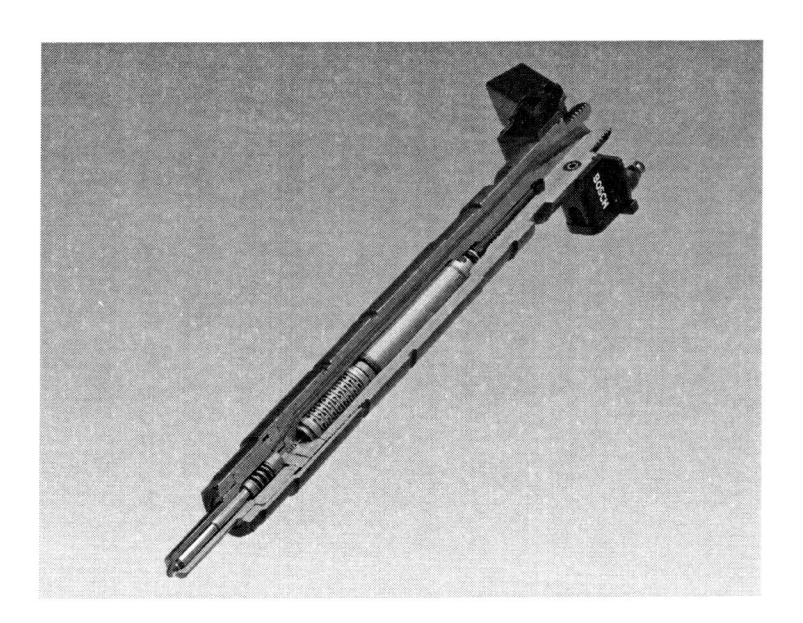

Figure 3.16 Bosch piezoelectric-actuated servo common rail injector.
(Courtesy of Bosch.)

Arguments can be made that the performance of solenoid-based injectors having a close-coupled servo mechanism is very comparable with piezo-based injectors coupled with a high-mounted servo mechanism. Today, except at idle, diesel combustion is essentially an open-loop activity, but the new generation of fast-response control actuators opens the prospect of closing the loop based on direct cylinder pressure feedback or other combustion indication [Ref. 3.8].

Whereas solenoid actuators have a force versus travel curve that is essentially nonlinear, the corresponding curve for a piezoactuator is both linear and responsive to actuation voltage. That is, at a given displacement, the force exerted is proportional to drive voltage. In general, the stack length determines the actuator axial displacement, and the material volume determines the available force (for a given voltage). To reduce drive voltage levels to relatively safe, nonlethal levels, multiple small piezoelectric transducers are stacked mechanically in series and are connected electrically in parallel, as illustrated in Figure 3.17.

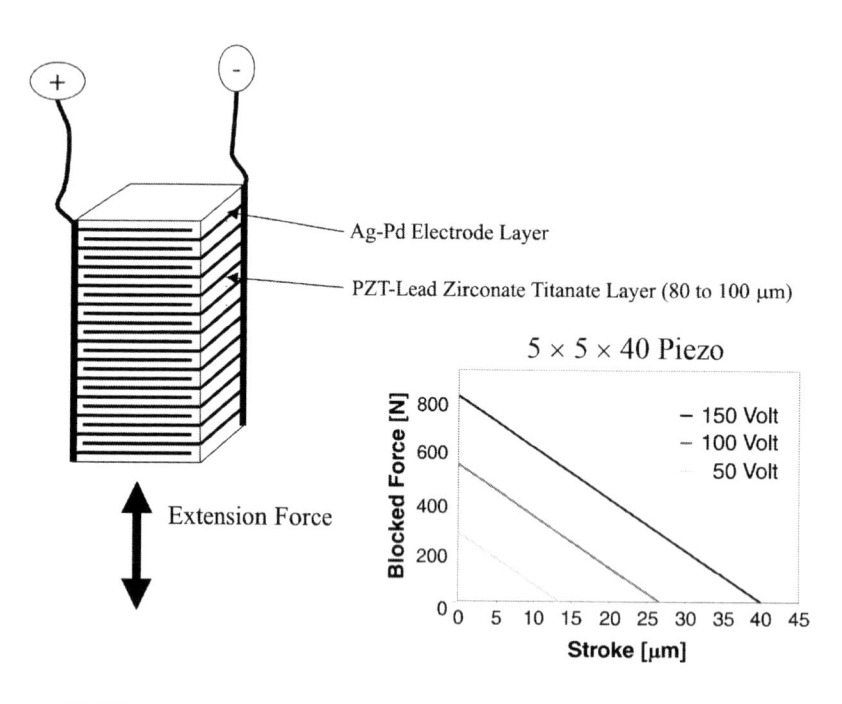

Figure 3.17 The construction and characteristics of piezoceramic actuators.

Direct Needle Actuation

As noted in Ref. 3.5, Hanover University has shown the theoretical potential of direct needle actuation when connected to a piezoceramic actuator. Also, in May 2004, Delphi announced that its next-generation common rail injector currently under development will eliminate the servo mechanism and feature direct actuation, too (Figure 3.18). This move brings several benefits that are likely to make this injector the new benchmark among common rail products. First, the needle velocity during the opening and closing transitions, at >2.5 m/s is significantly faster than for servo-controlled nozzles, for which 1 m/s is typical. This implies that the mean injection rate is higher for a direct-acting injector because the temporal rate shape is more nearly square. Additionally for a given injected fuel quantity, better atomization is obtained because less time is spent in the needle seat-throttling regime. This performance is achieved without compromise to the control of minimum fuel quantities; indeed, it is claimed to be more stable and has better resolution than servo-controlled injectors. Second, the construction of this injector is such that there is no return flow from the injector. Apart from the savings in low-pressure backleak plumbing on the engine, this also means that there is significantly less high-pressure leakage from the system. As such, it is expected to be able to operate at 2000-bar rail pressure much more efficiently than servo injectors.

Figure 3.18 Delphi direct-acting piezo-injector [Ref. 3.3].

The improved precision of control of this direct-acting injector results in better shot-to-shot performance, less interaction from one close injection to the next, and the ability to merge one injection with the next without a minimum separation constraint. This capability is enhanced by the potential for proportional needle lift control, which may prove essential for practical variable area nozzle realization. Put it all together, and this technology is expected to demonstrate the potential for a step-change improvement in emissions, depending on how the calibration is slanted. The improved control of multiple injections relative to a servo injector permits better matching of the "digitized" rate shape to the ideal profile, and the high-pressure capability with fast end-of-injection characteristics can be deployed to improve smoke and PM emissions.

As control valve actuator performance evolves, moving from top-mounted solenoid with unbalanced valve to internal piezoelectric actuator close-coupled to the nozzle, the possible number of discrete injections per cycle increases (Figure 3.19). First-generation common rail offered a practical limit of three injections, with limitations on minimum pilot quantity and pilot-to-main and main-to-post separation. The high dynamic mass of some systems imposed nozzle seat durability constraints based on the additional seat impacts incurred. Current and future systems offer vastly improved performance due to much lower dynamic mass, with, for example, Bosch claiming that the moving mass of its third-generation injector with piezoelectric servo mechanism is down to

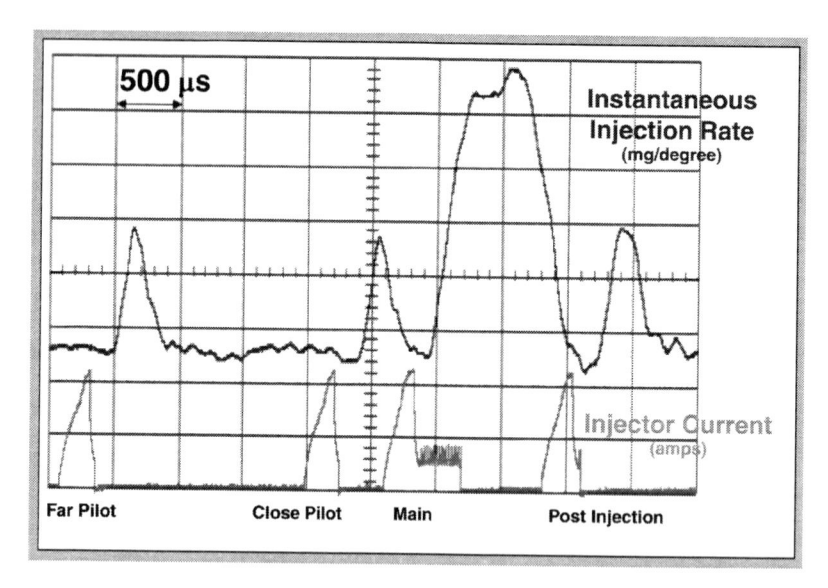

Figure 3.19 (Upper) Rate of injection and (lower) solenoid actuator drive waveforms for a multiple-injection event.

4 grams, from the 16 grams of earlier solenoid servo generations. Five injections over a normal diesel combustion event seems realistic, but potentially when early injections for homogeneous charge compression ignition (HCCI) and late injections for aftertreatment regeneration are considered, the number of injections per engine cycle could increase significantly.

Best-in-class common rail injection systems must excel in a number of performance-related areas if the host engine is to meet critical customer acclaim. Combustion noise is closely related to pilot injection capability, for which the challenge is to consistently deliver exceedingly small fuel quantities, frequently less than 1 mm^3/st. Depending on the engine speed, swirl level, and other operating conditions, the timing of this injection is critical; thus, the minimum separation between pilot and main injection or main and post can be a decisive factor. Additionally, some common rail systems offer superior idle-quality refinement by pilot fuel quantity feedback based on a block-mounted accelerometer signal. By "listening" to the structure excitation generated by combustion from a pilot injection and then applying an offset to the pilot pulse width on all cylinders individually to match a target excitation level, any drift in injector performance over time can be identified and corrected. This strategy permits consistent idle quality performance over the vehicle lifetime, and extensions of this technology can be used to minimize injector-to-injector delivery variation in the normal operating range.

Electronic Control Unit

The array of sensors and actuators associated with common rail systems is not unlike that for other electronic FIE, with the obvious exception that there is feedback and control of rail pressure. As discussed, feedback is provided by the rail pressure sensor signal, and pressure control is provided primarily by inlet metering of fuel to the high-pressure pump but supplemented by other control actions for fine regulation. Because the common rail contains fuel at potentially lethal pressure, and because a leak or other malfunction of an injector could have severe consequences on engine durability, system safety is a major concern for software engineering. For each application, these safety concerns will be identified and addressed through the normal failure modes and effects analysis (FMEA) methodologies, and several system-specific mitigation strategies are available. Additionally, system integrity is essential, because in all modern electronically controlled injection systems, fueling control implies full-authority drive-by-wire control of engine torque, not dissimilar to gasoline engine electronic throttle control, for which complex signal plausibility overchecks are required.

System architecture can be nontraditional for modern diesel fuel systems, in that there is often separation between the part of the controller that is very fuel-system specific and the part that is relatively generic to any engine controller. Thus, there may be an engine-mounted injection control unit (ICU) remote from the ECU, but linked by a standardized interface protocol. Whether this split arrangement or an all-in-one ECU is adopted, the precision of injection control at the injector requires very close matching between the drive circuitry and the actuator, much more so than is the case for gasoline injectors. This mitigates against the straightforward adaptation of a spark ignition (SI) controller. Additionally, there are usually some unique diesel-only features for which provision must be made, such as glow-plug control, supplementary block heater control, supplementary cabin heat control, and perhaps exhaust aftertreatment controls. This latter can be quite extensive, embracing numerous sensors such as NOx, catalyst temperature, exhaust ΔP, aqueous urea level and temperature, ammonia slip, and potentially particulate matter (PM) indication. The ECU outputs are likely to include drivers for the urea selective catalytic reduction (SCR) dosing and/or the hydrocarbon lean NOx trap (LNT) dosing system, as well as several control actuators particularly for the LNT system.

Control of exhaust gas aftertreatment systems will bring its own array of challenges. Depending on the storage capacity of the catalyst, SCR NOx reduction is a near real-time process, in which a fine balance must be constantly maintained between engine NOx production and the aqueous urea dosing rate. Too much urea will result in ammonia slip; too little, and NOx emissions will be higher than intended. An accurate NOx model or estimation methodology is essential in this case. Likewise, NOx adsorbers (LNT) require a rich regeneration every 60 to 90 seconds, which, depending on the system architecture, is likely to involve intrusive intervention to the steady-state engine calibration. Manipulating the many engine parameters such as EGR rate, variable geometry turbocharger (VGT) position, fueling, and injection timing to achieve this lean/rich/lean switch in a robust manner that is imperceptible to the driver under all speed and load conditions is a controls challenge that probably is best handled by a model-based strategy. In contrast, infrequent in-cylinder post injections for diesel particulate filter (DPF) exothermic regeneration is a relatively straightforward controls proposition.

Onboard diagnostics (OBD) to detect combustion and emissions systems defects over the life of the vehicle is an area that will become more demanding as the emissions limits for both failed and nonfailed conditions are reduced. The challenge is in making the calibration of the system malfunction diagnostic monitors sufficiently robust so that unintended malfunction indications are not flagged. Again, model-based controls may offer the logical solution. An extensive and complex test program involving design of experiments is usually required to evaluate all potential failure modes and system interaction effects to build confidence that exhaust emissions will remain within the proscribed limits. Although OBD requirements differ among the various regulating bodies, there are ongoing efforts at world harmonization, and it is beyond the scope of this work to enumerate the differences. Nevertheless, taking the California Air Resources Board (CARB) diesel OBD regulations as an example, the fuel injection system parameters that must be monitored include the following:

- Rail pressure in common rail systems
- Over-fueling errors
- Injection timing errors
- Multiple injection errors (e.g., a single injection where a multiple injection was intended)
- Misfire detection

The proliferation of injection events resulting from HCCI and post-injection strategies will result in overlapping injection demands from different cylinders, and this will impact the control valve drive electronics, not only in the number of individual drivers required but also in the heat dissipation that must be accommodated. This comment may be particularly true as more applications transition to piezoelectric actuators and as the cylinder count increases from four to six to eight. Then, as greater integration among engine subsystems such as EGR, VGT, and variable valve actuation (VVA) becomes necessary, optimizing the engine calibration will move from being merely challenging to a level beyond that. In turn, this trend has driven the move toward physical model-based control strategies and away from map-based control for the highly interactive parameters that otherwise are difficult to optimize under transient conditions. The advent of high-frequency response combustion feedback sensors, coupled with responsive injection systems and fast control processors, suggest the possibility of real-time NOx estimation for exhaust gas aftertreatment control. Thus, future diesel engine ECUs are likely to migrate to computationally intensive model-based control strategies, running on 32-bit processors that will likely include integrated digital signal processors (DSPs) with many channels for high-performance analog-to-digital conversion.

In terms of the control complexity, however, advanced FIE systems of any type offer many more degrees of freedom to the calibrator than earlier systems provided, and both competitive demands and emissions compliance requirements imply that all available features will be exploited. Moving forward, this situation will only grow in complexity as new features such as variable area nozzles with their essentially nonlinear or dual-mode gain curve characteristics become available. From the foregoing comments, it will be appreciated that arriving at an optimum calibration for an engine where there are so many interactive parameters is problematic, made more so by the narrow gates that must be navigated to satisfy the emissions standards. Additionally, in the future, it will not be enough to simply meet the standards under the closely controlled emissions test conditions of ambient temperature, barometric pressure, and fuel quality. It is also necessary to remain within specified limits for the not-to-exceed (NTE) requirements and the off-cycle standards (Figure 3.20).

The latter is of particular concern to engine suppliers in the heavy-duty sector, because responsibility extends from engine pass-off test, through the variations inherent in the multitude of on-highway applications, and across the national variation of fuel properties, to embrace ambient conditions between sea level and 1700 meters altitude. Compliance of the in-service fleet will be policed by the regulatory authorities through the deployment of portable emissions testing equipment.

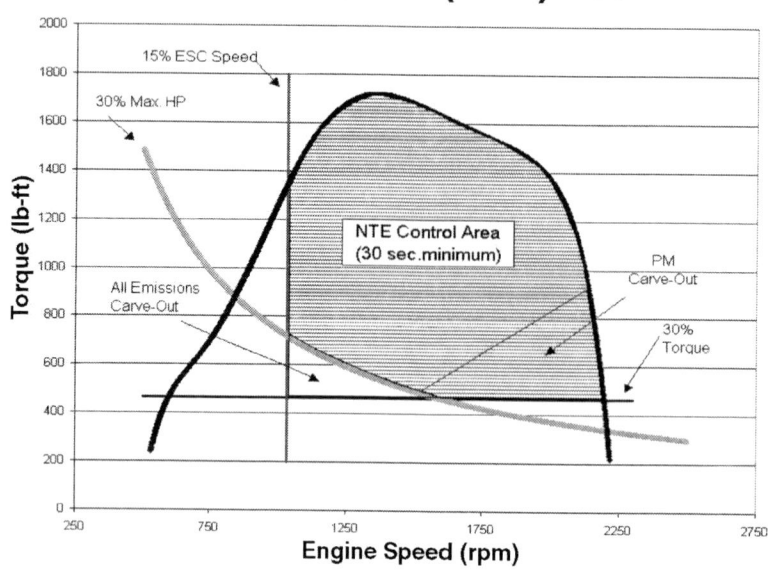

Figure 3.20 Illustration of an engine map "not-to-exceed" (NTE) control area.

Therefore, calibration robustness is of prime importance, and the difficulties of achieving this with the multiple variables involved has encouraged the development of automated mapping and optimization techniques that with minimum human intervention are intended to balance conflicting requirements between, say, NOx, PM, BSFC, peak cylinder pressure, turbocharger speed, and engine torque, to arrive at the best compromise. A reduction in test time and expensive test facilities for calibration activities might be assumed to flow from this capability once it becomes available. But currently, the various statistically based techniques have not demonstrated the necessary maturity for a decisive view on this point to be taken [Ref. 3.9].

Chapter Four

Hydraulic Electronic Unit Injectors

During the late 1970s and early 1980s, there was recognition that the then current fuel injection systems were a limitation in achieving growth in power density, particularly for military applications. Higher injection pressure capability for faster heat release rates and smoke reduction was a major objective, but current pump-line-nozzle (PLN) injection systems were pressure limited to perhaps 750 bar. Military R&D contracts therefore funded the development of a number of expensive but capable prototype systems based on unit injector configurations that could achieve peak injection pressures in the 1000- to 1200-bar range. One technique used to achieve these pressures was the integration of a dual-diameter piston intensifier into the injector, whereby a relatively low pressure, such as might be provided by a conventional fuel injection equipment (FIE) pump, is applied to the large-diameter piston. The resultant force, in turn, is applied to the smaller-diameter pumping piston, which generates the injection pressure. With a 3:1 area ratio, a 400-bar low-pressure stage could, in theory, provide for a 1200-bar injection pressure. These systems, such as the Helios (Lucas) and UFIS (AMBAC), although too expensive to be commercial, showed that the combination of high-pressure capability with electronic control was a feasible and logical path forward [Ref. 4.1].

By 1985, the power density limitation of the Ricardo Comet-based divided-chamber engines was apparent; thus, the next generation of U.S. full-sized pickup truck engine would have to transition to direct injection format if the "advertised power" race were to continue. For this new generation of engine, an equally new high-pressure FIE concept was needed, because none of the conventional PLN systems was suitable. In the absence of any other appropriate system such as high-pressure common rail (HPCR), the logical product was a hydraulically intensified, electronically controlled unit injector (HEUI). A mechanically actuated electronic unit injector (EUI) was not appropriate, because the typical single vee-mounted camshaft had neither the space for the extra pumping lobes nor the "overhead" stiffness for generation of high pressures. To fill this need, Caterpillar Tractor Co. took the initiative to design and develop its HEUI product for the International Truck Co. 7.3-liter Powerstroke V8 as supplied to Ford Motor Co. [Ref. 4.2].

Unlike earlier systems, the Caterpillar HEUI FIE system uses the engine lubricating oil as the hydraulic fluid for the first (low-pressure) stage (Figure 4.1) [Ref. 4.3]. The decision to use lube oil rather than diesel fuel as the first-stage fluid permits some manufacturing tolerances to be relaxed, to the benefit of system cost, in exchange for cold flow viscosity and aeration challenges. The system comprises an engine-driven variable displacement multi-piston pump that raises the first-stage rail pressure to a controllable value in the range 60 to 280 bar, galleries that conduct that fluid to the individual injectors, and electronically controlled injectors having an integral pressure-intensifier piston. Engine fuel is supplied to the injector via other conduits in the cylinder head, entering the pumping chamber situated conveniently close to the nozzle. Depending on the area ratio existing between the intensifier piston and the pumping piston, usually being in the range of 5:1 to 7:1, the resulting injection pressure will have a direct relationship to the lube oil rail pressure.

The HEUI concept in the first generation provided for independent control of injection pressure through modulation of the intermediate rail pressure, and injection timing and quantity control via an electromagnetically actuated valve that admitted rail pressure to the intensifier piston. At first, only marginally commercial with respect to combustion noise in the full-sized pickup truck application, a mechanical pilot injection feature was soon added that gave a significant improvement. A second-generation HEUI has now been released, which, by adding sophistication to the control valve and actuator, provides multiple injection capability and some measure of rate-shaping to the injection profile (Figure 4.2) [Ref. 4.3].

In summary, HEUI as pioneered by Caterpillar met the market demand for a high-pressure electronically controlled injection system, suitable for the new generation of emissions-controlled direct injection (DI) medium-duty diesel engines. It offered installation advantages over the only other logical competitive system of mechanically actuated electronic unit injector (EUI, or MEUI), and the additional feature of speed-independent injection pressure control was expected to be a valuable emissions calibration asset. As such, it has been highly successful in its intended market but has not made significant inroads into markets on either side. In the light-duty market, HEUI injector bulk has been an application impediment, as too has been the perception that

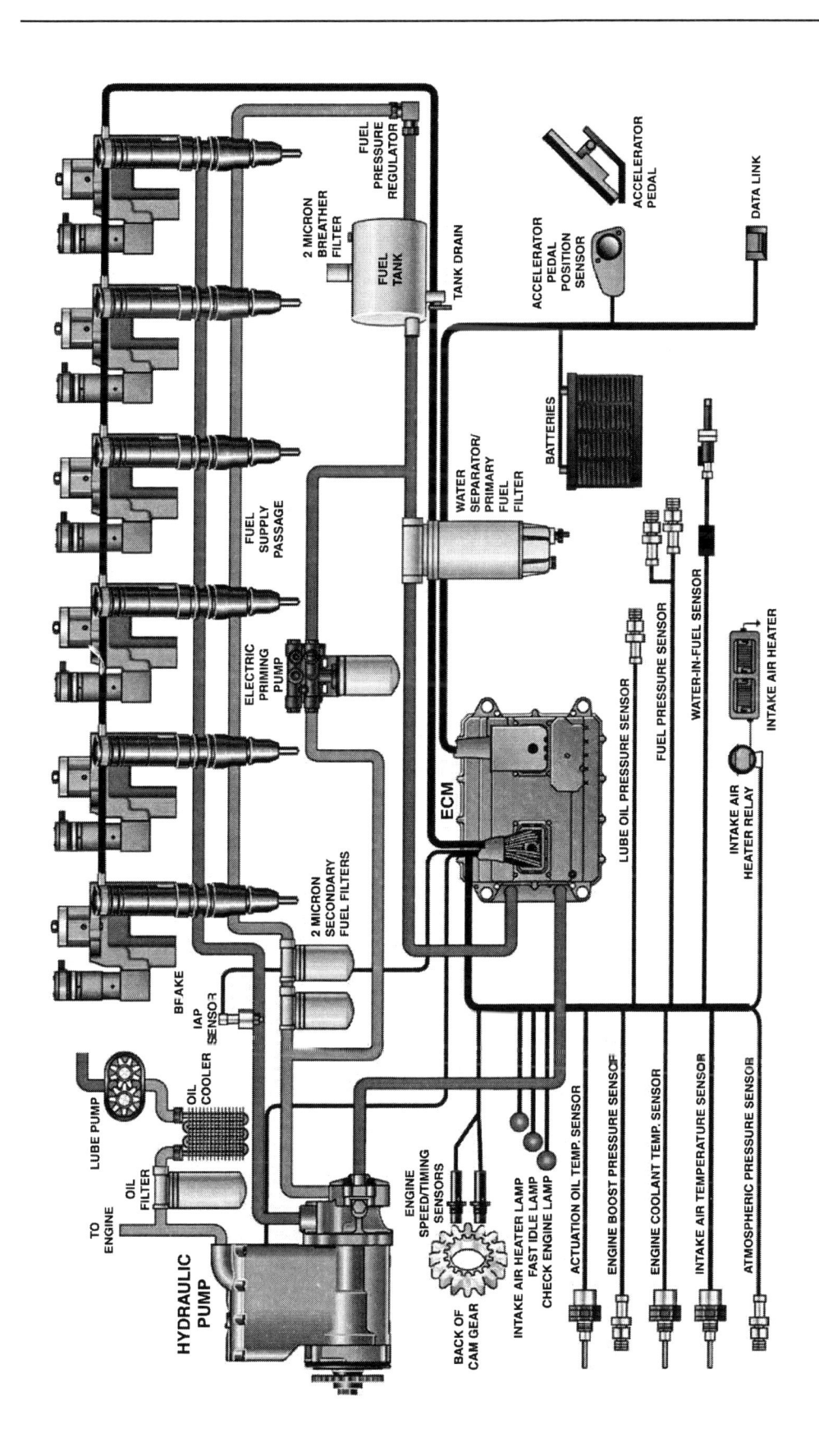

Figure 4.1 Caterpillar HEUI system schematic [Ref. 4.3].

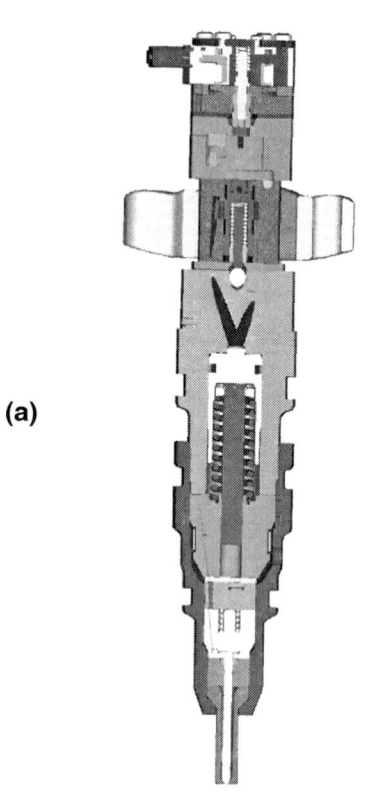

Figure 4.2 (a) Caterpillar HEUI-B injector, and (b) suggested rate-shape schedule [Ref. 4.3].

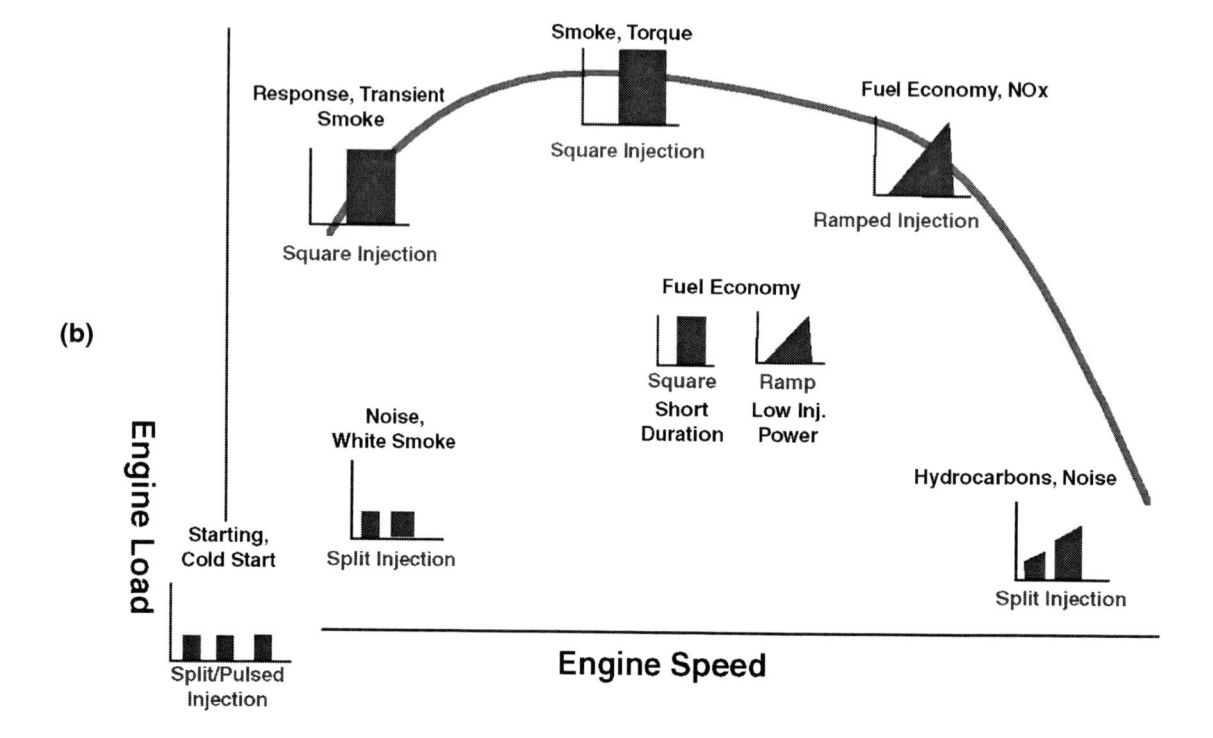

the injection precision and flexibility are not competitive with other high-pressure common rail systems. In the heavy-duty sector where high mechanical efficiency and low fuel consumption are valued, it has, so far, made slow inroads against the incumbent MEUI.

Another example of this type of fuel system is the Siemens G2 injection system in the Model Year 2003 6.0-liter V8 Ford Powerstroke engine. The G2 is the result of a joint venture between International Truck & Engine, Sturman Engine Systems, and Siemens, in which broadly International commissioned the product, Sturman designed and developed it, and Siemens is the manufacturer. Similar in concept and construction to the Caterpillar HEUI product, the G2 uses the novel double-acting "digital" actuator proprietary to Sturman described in Chapter Three, in which the spool of the control valve is also the solenoid armature.

While the G2 has similar limitations as HEUI in terms of application potential, they both have a fundamental opportunity for the synergistic integration of electrohydraulic valve actuation. This is the case because the fuel system already incorporates a large-capacity medium-pressure hydraulic pump, essentially similar to that required for the variable valve actuation (VVA) or variable valve timing (VVT) system. For a combined system, the pump will undoubtedly need to be of greater displacement, but the convenience of having the hydraulic energy source in a gallery or rail in the cylinder head adjacent to both the injector and valve actuators is obvious. One strategic issue that needs resolution is that while rail pressure currently varies over the speed and load envelope as a function of injection pressure calibration requirements, the same pressure map is unlikely to be optimum for the VVT system. Resolution of this dichotomy may require a compromise in the pressure map values to suit both systems, or possibly a ganged pump with two separate high-pressure galleries. Clearly, once electrohydraulic VVT becomes available, there is the opportunity to separate the engine fluid systems—"dirty" lube oil remaining in the crankcase, and "clean" hydraulic oil in the cylinder head, with different change intervals as appropriate.

Chapter Five

Unit Injectors and Unit Pumps

In the early 1990s, during the period when it was apparent that the light-duty diesel car market was here to stay but the search was on for a viable direct injection (DI) engine concept, it appeared as if the electronically controlled unit injector (EUI) was going to be the predominant fuel system. In comparison with indirect injection (IDI) combustion systems, DI requires much higher injection pressures, higher rates of injection, and greater precision in injection timing and hydraulic performance, all of which pushed rotary distributor pump-line-nozzle (PLN) systems to their limits. The only viable existing system that could meet the performance objectives was the EUI or its close cousin electronic unit pump (EUP), given that the choice at that time was between in-line pump, rotary distributor pump, and EUI. However, it did bring some negative aspects with it, principally issues of bulk and camshaft drive torsionals. Nevertheless, some engine OEMs who were contemplating new engine designs at that time gave it serious consideration. Among that select group were Ford, Land-Rover (Figure 5.1), and Volkswagen, with the latter two taking their engines into production circa 1996.

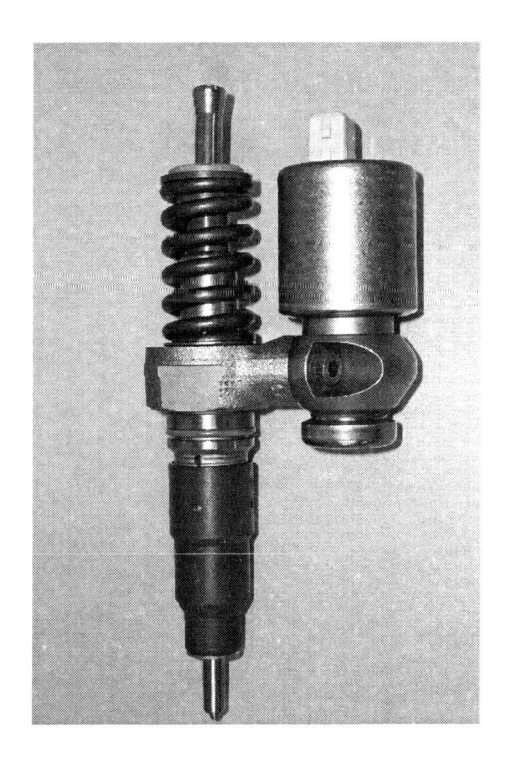

Figure 5.1 Electronic unit injector (EUI) for a 0.5-liter/cylinder engine.
(Courtesy of Delphi.)

At this point, it is appropriate to examine the key features of the EUI and to discuss its attributes and future development. As with other fuel injection equipment (FIE) concepts, unit injector history stretches back to the beginning of "solid" injection, but it became an established technology with its adoption in the 1930s when it was championed by the Winton, Murphy, Detroit-Diesel, and Cummins engine companies. As described earlier, the essential feature is that the pump and the atomizer are in a monoblock unit together and are located in the cylinder head, with one device per cylinder. The steel body houses both the pumping plunger and the nozzle assembly. Also, because of their close proximity, the total pressurized fluid volume is comparatively very low, which gives a crispness of performance generally free from reflected waves that is not available from other systems. This results in a high efficiency of pumping performance that can be expressed in the spread-over ratio—being the ratio of theoretical pumping rate derived from plunger area multiplied by cam velocity versus the actual measured pumping rate. It is much closer to unity for unit injectors than for PLN systems [Ref. 5.1].

Many of these comments apply equally to the EUP, which is, from one perspective, a throwback to conventional PLN FIE. Generally adopted for packaging reasons, the EUP separates the pump module from the injector with a high-pressure pipe, but the individual pump units are driven from the engine camshaft, which may be in the block or in the head, and historical examples of both have existed (Figure 5.2). Because these pumping elements are driven by a robust camshaft and may be placed conveniently close to their respective cylinder so that only a very short high-pressure pipe length is required, the loss in system stiffness relative to an EUI configuration can be held to a very low value. Advantages that the EUP system may confer to the total engine design include a reduction in engine overall height because an overhead camshaft is not necessary, and a less-complex cylinder head in which the injector claims less space. This results in a lower installed cost relative to EUI.

The low hydraulic dead volume of the EUI contributes to system stiffness, encouraging in turn the use of a stiff drive mechanism, both of which are necessary to support the high injection pressures of which unit injectors are capable. Unlike other FIE that can almost be added on as an afterthought (albeit at the engine developer's peril), unit injectors are much more closely integrated into the base-engine design, a fact having both pros and cons. Close liaison between the engine designer and the FIE supplier is necessary to arrive at an optimum integration, so that the engine designer provides a drive mechanism of sufficient stiffness (typically around 20,000 N/mm), and the injector configuration impinges minimally on cylinder head design or engine height. Achieving this level of drive stiffness is not easy and generally results in adoption of an overhead camshaft and rocker mechanism (Figure 5.3), although direct attack cam actuation (which is stiffer but introduces service problems) or short pushrod and rocker (which is less stiff) are alternatives. Careful finite element analysis (FEA) is necessary to arrive at an optimum solution. As injection pressures climb, adoption of nontraditional materials, such as ceramics, becomes justified in highly loaded force-path components such as the cam-follower roller.

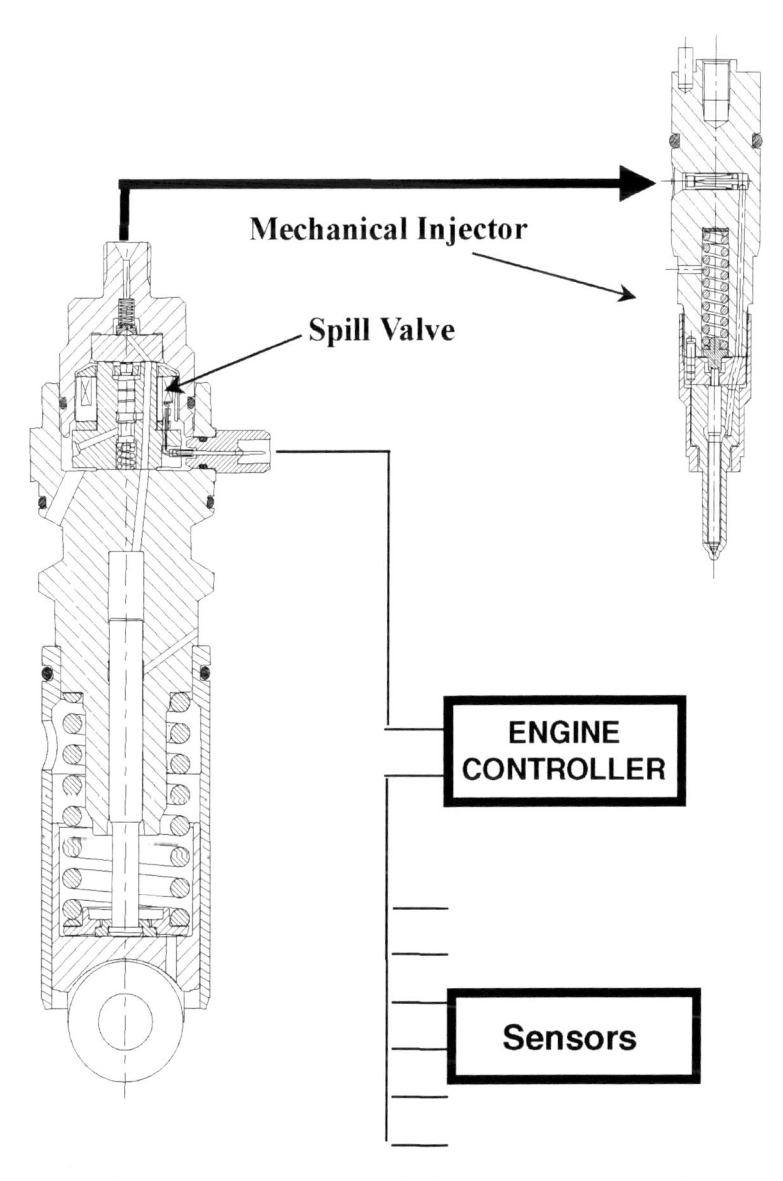

Figure 5.2 Electronic unit pump (EUP) and spring-controlled injector, with short high-pressure pipe. *(Courtesy of Delphi.)*

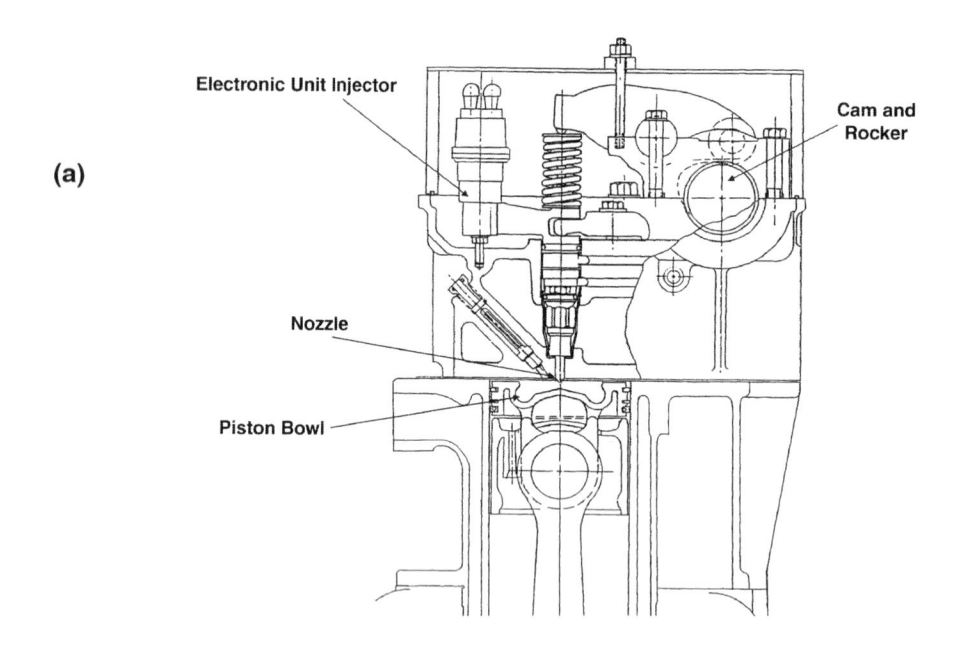

Figure 5.3 (a) Side view and (b) photograph of an EUI installation in a sectioned cylinder head.

In the high-mileage heavy-duty truck sector, fuel purchase represents a large proportion of the total operating costs; thus, engine concepts that deliver minimum real-world fuel consumption are favored. Because diesel combustion requires the careful and thorough mixing of fuel and air, that mixing energy must come from somewhere, either controlled air motion or fuel spray momentum, or both. The generation of air swirl and squish costs energy, as does the generation of fuel pressure. However, it has been found in practice that the best efficiency is obtained from quiescent (very low air motion) combustion systems, in which the mixing energy is provided by the fuel spray momentum alone and virtually none from the air motion. Therefore, this implies a requirement for very high injection pressures, for which the unit injector is the best available FIE system and explains its dominant position in this market sector.

Single-Valve Electronic Unit Injectors

Whereas traditional mechanical unit injectors have utilized a rack and plunger-helix mechanism for fuel quantity control, similar in concept to that used by in-line pumps, EUIs normally employ a solenoid or, less commonly, a piezoelectric-controlled "spill valve" for this purpose. The method of control, sometimes referred to as "finger-in-the-dike," makes use of a relatively simple normally open two-way valve to either contain the internal hydraulic pressure or permit it to spill back to the tank. In operation, therefore, as the cam-driven plunger descends, the displaced fuel below it has only the option of returning to the tank via the open spill valve. However, when injection is desired, the spill valve closes, and further plunger-displaced fuel is converted into pressure, which opens the nozzle against spring preload so that subsequent fuel is injected through the nozzle at an ever increasing pressure and rate. Opening the spill valve terminates injection through a rapid pressure collapse, which, if well managed, should fall at a rate of around 2000 bar/ms or better. Whereas helix-controlled mechanical unit injectors were limited in their injection pressure potential due to seizure propensity from plunger sideload, EUIs employ plain plungers that have a significantly higher pressure capability. Thus, peak injection pressure capability is normally determined by the structural robustness of other components in the system, ranging from the cam gear drive, cam-roller hertzian stress, or injection nozzle strength.

Therefore, with the single-valve EUI, the start of injection is determined by when the spill valve closes, and the fuel quantity injected by the duration of closure. As noted with EUI, because the spread-over ratio is close to unity, a good relationship exists between geometric plunger displacement and actual fuel delivered. This makes fuel control through the engine control unit (ECU) relatively straightforward in comparison to nonlinear systems. Behind this, though, it must be appreciated that injection events are constrained to occur only when a cam "bump" is available and not everywhere within the crank-angle domain as with common rail. For conventional diesel combustion, this constraint is not normally a problem because injection and combustion always fall within a small window close to top dead center (TDC).

But flexibility is the keyword for the future, and this narrow timing constraint is no longer acceptable. New mixed-mode emissions-compliant low-temperature combustion strategies such as homogeneous charge compression ignition (HCCI) demand the ability to inject fuel at hitherto unforeseen crank angles, such as during the intake and early compression cycles. Additionally, single-valve EUI was eclipsed by common rail in two other critical areas: pressure control and multiple injection capability. We have seen with common rail FIE where control of rail pressure is used as a fundamental analog for injection rate, and this is proving to be an essential calibration variable. No such variable existed for EUI because the injection rate is so closely related to cam velocity, and the injection pressure rose with engine speed as for other cam-driven FIE. Likewise, single-valve EUI is capable of only rudimentary pilot injection and not the sophisticated multiple injections of common rail.

Two-Valve Electronic Unit Injectors

To level the playing field again between common rail and EUI, the latter system has moved to a two-valve strategy, in which one valve controls injection pressure and the second valve governs injection timing, fuel quantity, and the multiple injection capability (Figure 5.4). In the case of EUP, this second valve resides, of necessity, in the injector. This broad range of functionality may be commanded ideally by two separate actuators, one per valve as exemplified by the Delphi E3 injector [Ref. 5.2], but it is possible to manage it with one actuator, as shown by the Caterpillar HEUI-B, with some loss of capability at the margins. Introduced to the heavy-duty sector initially where it has been recognized as a key requirement for U.S. 2007 complaint engines, this technology is expected to appear in the medium- and light-duty unit injector market in due course.

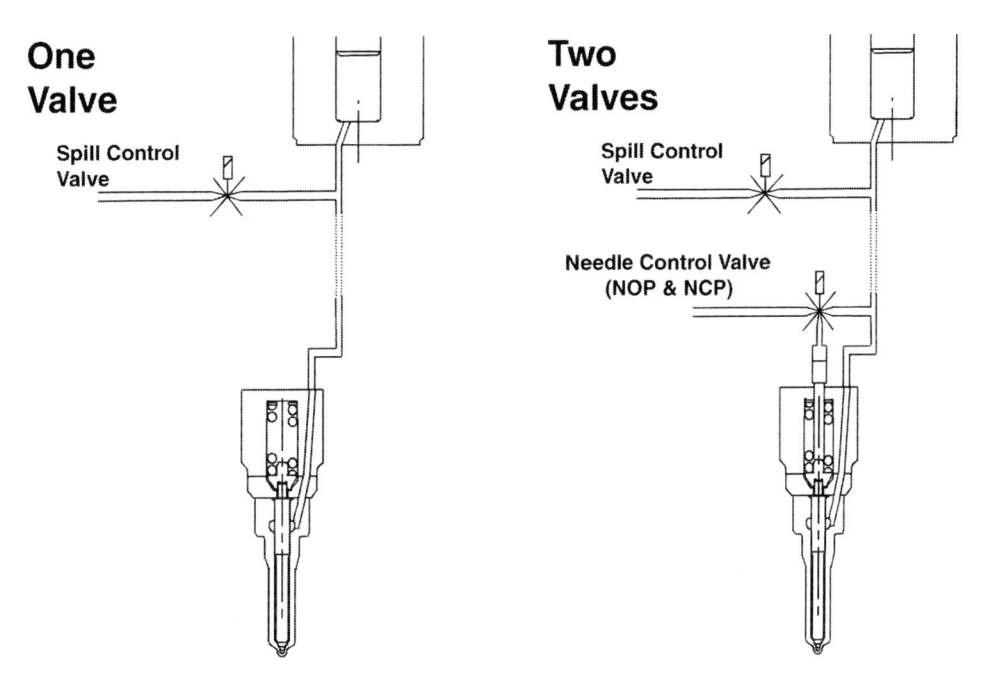

Figure 5.4 Single- and two-valve EUI schematic [Ref. 5.2]. The needle control valve (NCV) governs the nozzle opening pressure (NOP) and nozzle closing pressure (NCP).

Operation of the two-valve EUI can be hard to grasp in comparison to common rail because there are so many degrees of freedom involved, with complex interactions between each. Taking as our example the Delphi E3 injector, there is a conventional two-way spill valve and a three-way needle control valve, both internal to the injector, located between the plunger and nozzle. As it is for the single-valve EUI, the spill valve determines when pressure is allowed to build inside the injector. Closing the valve sooner results in a higher internal pressure at the appointed time for start of injection than if a later spill-valve closing is chosen. Thus, a measure of pressure control is available.

The needle control valve (NCV) is a three-way valve that, when in its de-energized (default) position, allows the internal pressure being built up within the injector to connect with a bias piston chamber, the piston of which adds additional load to the nozzle needle in the closing direction. Thus, the needle remains on its seat, irrespective of the magnitude of the internal pressure because the nozzle spring load plus bias load far exceed the opposing (opening) force of the pressure acting on the needle differential area. Only when the needle control valve is switched, at which point the bias piston chamber is first isolated from the internal pressure and then connected to drain, does the force balance change. Now the nozzle opens at the pressure that exists within the injector at that moment, which might be anywhere between 300 bar and, say, 1800 bar. In this manner, shot-to-shot control over nozzle opening pressure and therefore injection rate is enabled. That is, a low nozzle opening pressure will provide a rising ramp rate profile; a high nozzle-opening-pressure selection will deliver a square profile or even a falling rate, which are typically best for high exhaust gas recirculation (EGR) conditions.

Rapid cycling of the needle control valve will provide options of a split injection, a close pilot and a main, a main and a close post, or pilot-main-post injections all within the normal combustion event. The natural behavior of the system, in contrast to common rail, is to make each successive injection within a multiple set at a higher pressure than the preceding one, as indicated in Figure 5.5. According to generally accepted combustion theory, this is beneficial because a low-pressure pilot and a high-pressure main and post injections are generally desirable. Likewise, fast termination of injection also is considered essential for minimum smoke emission. In this case, rapid needle seating is assured when the needle control valve is de-energized, under the action of both the needle return spring and reestablishment of the force bias pressure.

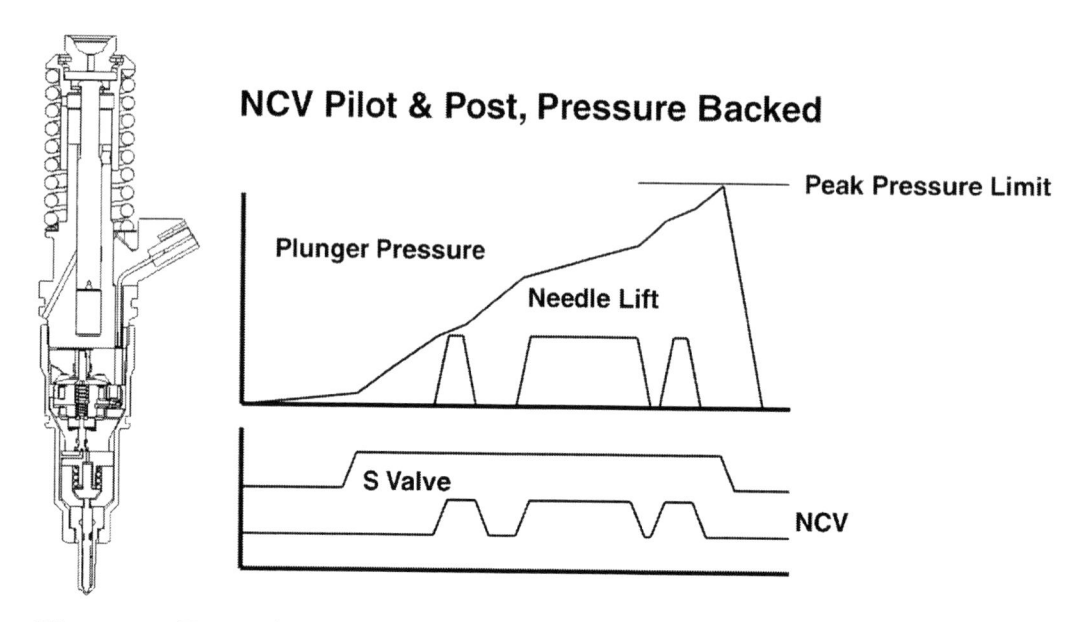

Figure 5.5 Two-valve EUI, showing the relationship of pressure and injection to control signals [Ref. 5.2].

Further flexibility is provided if the needle control valve is energized prior to the spill valve, because in this instance, the two-valve injector is able to respond in the same way as a single-valve type, providing spring-controlled (i.e., low-injection-rate) nozzle opening, and, if desired, spring-controlled termination. Whichever mode is selected, it is normal to arrange for a near-synchronized ending in which both actuators (the spill valve and the needle control valve) are de-energized almost together, concurrent with the end of the final injection event.

Summary of Electronic Unit Injectors

Two-valve EUI brings to the marketplace many of the benefits of common rail, plus some pros and cons of its own. As noted, it can deliver injection pressure control and multiple injections, but neither is quite as flexible as is available from common rail FIE. Peak injection pressure is largely decoupled from speed effects and brings the advantages of a higher absolute value than current common rail systems can provide, being typically in the range of 2000 to 2500 bar. Moreover, unlike common rail, which requires several engine revolutions to change injection pressure, a different pressure can be commanded on a cycle-by-cycle basis—a facility that can be very helpful during transient maneuvers. Further, in a close multiple injection group (say, a pilot, main, and post injection), there will be an ascending trend to the peak pressure and therefore injection rates over the total event—that is, a low injection rate for pilot and ending in a high-rate post. Again, this characteristic is not available from current common rail FIE, where all injections occur at the same compromise rail pressure. Thus, it could be said that the EUI represents a leading candidate FIE option for diesel engine applications that desire to challenge the spark ignition (SI) engine in specific power rating.

On the debit side, there is less flexibility with EUI in the provision of additional injection events away from the conventional timing point, as may be required by HCCI or late-post injection for exhaust aftertreatment regeneration. Currently, to address these requirements, additional appropriately timed cam "bumps" may have to be considered. Additionally, EUI, but not EUP, carries the penalty of a relatively bulky injector in the cylinder head, and a robust cam drive and rocker mechanism that bring difficult noise, vibration, and harshness (NVH) problems to the design. These do not get easier as injection pressures increase and can require solutions such as pendulum dampers in the drive gear to smooth out the impulsive drive loads.

Chapter Six

Current Application Issues

Fuel Injection Equipment Selection

Fuel injection equipment (FIE) control is at the heart of diesel engine combustion strategy. In direct injection (DI) combustion systems, the diesel spray characteristics dominate the air-fuel mixture formation, as well as ignition, combustion, and pollutant formation processes. Moreover, these characteristics must be balanced over a wide dynamic range of speed and load, and under different ambient temperatures. The requirements placed on the fuel injection system for different classes of engine also vary. For example, in small-bore high-speed DI engines, special attention to liquid-phase fuel penetration is required. In the interest of reducing particulate matter (PM) and hydrocarbon (HC) emissions, care must be taken to avoid the impingement of liquid-phase spray onto the combustion chamber walls—an especially difficult task at low-speed and low-load conditions. At high speed and high load, careful matching of the sprays in terms of number, orifice diameter, and spray cone angle with the combustion chamber shape is necessary to minimize smoke emissions and to maximize air utilization.

At the time of this writing (2005), there are at least seven or eight manufacturers of common rail FIE across the different market sectors, although not all are of the high-pressure common rail (HPCR) type. Slightly fewer are in the electronic unit injector (EUI) or electronic unit pump (EUP) market, and perhaps two are in the hydraulic electronic unit injector (HEUI) market. All of these systems have been described in more detail elsewhere in this work. In terms of production volume, common rail is, without question, the predominant system for the foreseeable future. Nevertheless, it is necessary to understand the advantages and disadvantages of the various system concepts and to make a well-considered choice in the selection of fuel system for a new engine program. Each system has its pros and cons, and it is essential to match the system attributes with the application requirements, both in the short term and in respect to long-term product evolution. The multiple drivers of emissions legislation, electronics capability, manufacturing technology, and customer expectations are, at the moment, combining to push rapid technological advance in the diesel engine and FIE markets, such that it is easy for the development road maps of engine and FIE to diverge if an uninformed decision is made early.

From a first-order perspective, it would appear that in the automotive field, HPCR is the optimum choice for light-duty applications, HEUI for medium-duty applications, and EUI or EUP for heavy-duty engines. However, looking beyond that paradigm, there is another that suggests that HPCR is the choice for refinement, and EUI is the choice for high specific power output. The reasons for that are explained in the respective chapters, but the point is that it is unwise to set those beliefs in stone, because both products are evolving rapidly, and the balance of attributes will change with time.

Injector Installation

Finding adequate space for injector installation has always been an issue for the engine designer. The first requirement has always been to maximize valve area and thus engine breathing ability, and then to provide sufficient cooling between the valve bridges. Too often, the space remaining for injector installation has required injector compromise to the detriment of system performance. Due to the exigencies of meeting emissions reduction mandates, this design ethos has been replaced of late by an appreciation for the need to optimize the combustion process using all available methods. In turn, this has driven an approach in which providing for the combustion system requirements has assumed the highest priority. The predominant combustion system for new emissions-controlled engines today is the four-valve, central vertical injector, DI reentrant bowl-in-piston arrangement, which, although certainly far from new in the heavy-duty field, has now become the norm in the light-duty field also (Figure 6.1).

Historically, the injector manufacturer has always been under pressure to make the smallest product possible, particularly with respect to body diameter, because as a supplier, the manufacturer is obliged to respond to the engine designers' demands. As a result, the basic outline dimensions of injectors have become standardized on a few specific diameters, the majority of which subsequently have become encoded in one or more ISO standards. For automotive applications, the popular body diameters in the past have been Ø21, Ø19, Ø17, Ø14, and Ø9.5 mm, the latter example generally being referred to as a pencil injector. From the engine designers' perspective, where they have been able to deliver the required durability, the Ø14- and Ø9.5-mm

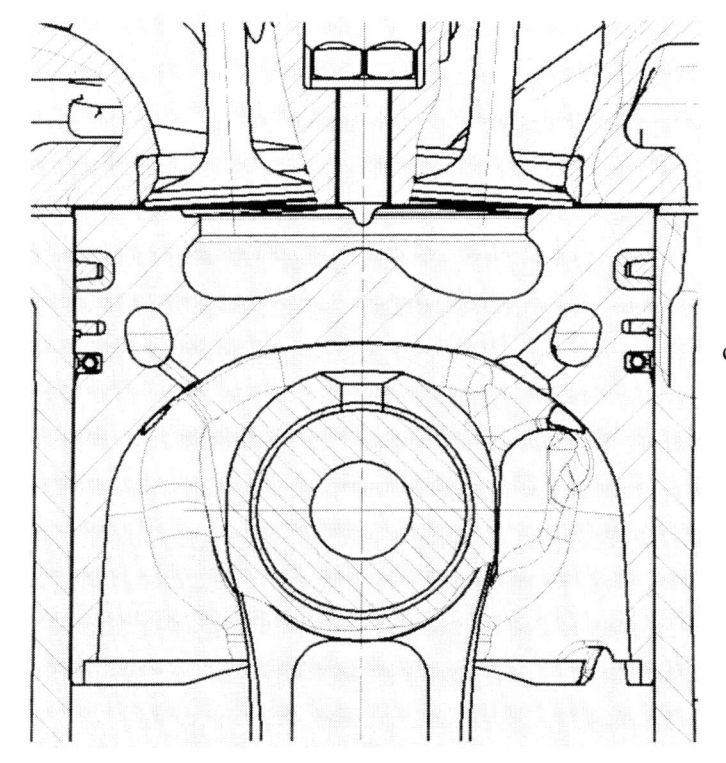

Figure 6.1 Four-valve direct injection (DI) combustion chamber with central vertical injector showing a reentrant combustion bowl; BMW four-cylinder diesel motor.

sizes have always been attractive. However, with the emphasis on emissions performance and associated durability today, the smallest practical body is Ø17 mm. Injectors so dimensioned have been able to deliver the structural robustness at high pressure ratings and to meet the other functional performance criteria demanded today.

In addition to minimizing injector body diameter, it is desirable to minimize the nozzle diameter. Small internal diameters within the nozzle, consistent with adequate flow areas, help to reduce the hoop stresses that otherwise may be problematic as injection pressures inevitably rise. Further, it is highly desirable to minimize the face area of the nozzle directly exposed to the combustion chamber, for easier thermal management of the nozzle. Combustion heat transferred into the nozzle is obliged to travel the length of the nozzle stem before it can be conducted into the cooled cylinder head at the injector seat interface. For these reasons, normal nozzle stem diameters have settled on either Ø9 or Ø7 mm, with a strong trend in favor of the latter. As specific power ratings increase in the future, unwelcome thermal issues with the nozzle, such as internal fuel lacquering, are likely to assume greater importance. Where uncooled cylinders are contemplated, such as may be the case for ceramic or adiabatic engines, it is likely that independently cooled nozzles must be specified. In this case, cooling channels are machined into the outer diameter of the nozzle stem, through which fuel, air, or other coolant is circulated, the stem then being encased in a thin-wall sheath, which may bring its diameter up to Ø10 mm or more. Separate coolant drillings in the injector body communicating with the nozzle are required, as is a dedicated external cooling circuit.

Another area of compromise is that of nozzle stem length. On one hand, the engine designer would like a long stem because that makes it easier to provide large ports and cooling passages in the cylinder head if the transition from the Ø7-mm stem to the Ø17-mm body is made as high as possible. On the other hand, difficulty of nozzle manufacture is exacerbated as stem length is increased. Dimensional concentricity of the internal needle seat in the nozzle body is a critical and closely held parameter, upon which acceptable light-load spray performance depends. Achieving the necessary concentricity at some distance down a blind bore with no local guide diameter and in high production volumes is a serious manufacturing challenge. The compromise between the desire for long nozzle stem length and close-tolerance seat concentricity typically results in a stem length of approximately 22 mm from the injector body seating face to the nozzle tip.

The constraint imposed on cylinder head design by a Ø17-mm injector body diameter and a 22-mm nozzle stem length in terms of port breathing capacity and valve bridge cooling may impose a lower limit on cylinder bore size for which the conventional injector is suited. Certainly, the cylinder head of small-bore automotive engines is crowded, particularly when a glow-plug starting aid is included. In the overriding interest of minimal surface-to-volume ratio and crevice volume, current engines normally have vertically disposed valves that encroach on space that the injector might otherwise claim. However, some recent small-engine cylinder head designs have been forced to cant the valves apart slightly, so that the injector may be accommodated between them. This stratagem will result in small valve-relief pockets in the fire-deck, with a corresponding increase of dead unusable air; therefore, it is not a preferred solution, particularly in engines of small cylinder capacity where such air can assume a large percentage of the total.

Injector Retention

Two common methods for clamping the injector into the cylinder head are through the medium of a tubular gland nut, or with a "crow's foot" clamp. The objective must be to provide a reliable clamp load that holds the injector squarely onto its seat with a force that is significantly greater than the maximum combustion pressure that will oppose it. Generally, the injector will seat onto an annealed copper washer where a flat seat is used, but some installations use a conical seat, particularly on large injector sizes. However, a conical seat design can result in greater variation of nozzle tip protrusion, a parameter for which it is important to check the sensitivity of the combustion system. In some installations, the injector clamp loading also serves to maintain integrity of the injector assembly, which otherwise is subject to high separation forces due to the internal hydraulic pressure. Thus, relaxation of the specified clamping load can result in combustion gas blowby past the injector seat or, in some cases, fuel leakage from the injector itself (Figure 6.2).

Fuel System Test and Evaluation

Mention has been made of some of the many performance-related parameters that are important in a modern diesel fuel system, but a recap would be appropriate at this point from the perspective of a typical development

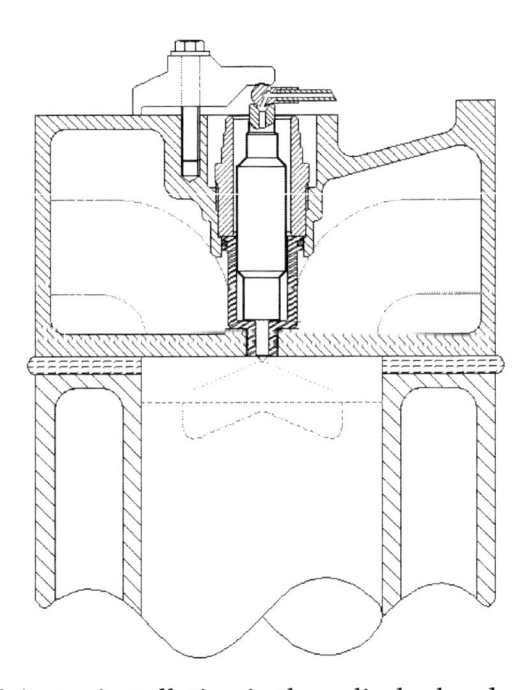

Figure 6.2 Injector installation in the cylinder head, with combined fuel feed connection and axial clamp loading.

program. With respect to the common rail pump, one is typically interested in the drive torque signature; noise and fuel temperature rise across the speed, load, and pressure map; and the transfer pump suction and flow characteristics. In the case of the injector, many more facets of interest must be measured, for which specialized equipment often is available. Some of the parameters are obvious, such as injected fuel quantity, injection (or rail) pressure, backleak flow, and electrical drive characteristics for the control valve. However, several other aspects are either less obvious or at least harder to measure, and these include the individual fuel mass amounts in close-coupled pilot, main, and post injections and their instantaneous injection rates. Beyond that, there is the injector delivery shot-to-shot consistency, and the same for injector-to-injector performance. Additionally, an indication of injector needle-lift can provide valuable information for understanding the combustion event. In some injector configurations, it is difficult to measure the injection pressure at or close to the nozzle; in those situations, it is usual to infer the pressure from some other parameter. As an example, for unit injectors, the injection pressure often is derived from the actuating rocker strain measurement, because this is a good, accessible analog for the desired signal.

When it comes to the nozzle characteristics, some specialized laser-based equipment is available for detailed examination of the spray structure, liquid-phase penetration, and droplet mass-time history. Other less esoteric criteria such as nozzle static flow, hole-to-hole flow, spray patternation, and spray momentum can be quantified with equipment of somewhat less sophistication. In diesel FIE application development, it is usual to conduct much of the early systems work on one of the various proprietary test benches that, to an extent, simulate the engine. They provide a variable-speed drive for the pump and the necessary fuel supply, collection, and flow measurement systems required for the convenient characterization of the FIE. On such a machine, it is possible to exercise the fuel system as if it were on engine, but with the advantage that almost all of the important parameters other than actual combustion can be readily accessed for measurement. An example is shown in Figures 6.3 and 6.4.

Figure 6.3 Common rail system installed on test bench. *(Courtesy of Delphi.)*

Figure 6.4 Fully automated test bench.

Chapter Seven

Future Outlook and Technology Trends

Looking ahead, some things at present seem immutable: the continued demand for higher injection pressure capability where direct injection (DI) combustion systems are concerned, and a desire for greater precision and flexibility of control over the injection event. Continued effort to seek the optimum technology balance based on total system cost between engine-out emissions reduction measures and the complexities of exhaust aftertreatment can be assumed. Different market sectors are likely to arrive at somewhat different answers, but it is important to get it right for the long term because this current wave of development directed toward the automotive diesel engine may be the last major effort before available investment dollars are redirected into fuel cells or other powertrain technology. Diesel engines are at an appreciable cost disadvantage with respect to spark ignition (SI) engines, most noticeably in the light-duty sector, with no obvious prospect of changing that relationship. Indeed, the additional technology required to make diesel emissions compliant in the future means that cost-reduction pressures will be a continuing fact of life for the engine development manager.

We can assume that ongoing pressure for exhaust emissions reduction, fuel consumption reduction, and specific power growth will remain as primary drivers into the future. This is likely to remain true, even as the cost effectiveness of exhaust aftertreatment improves, or new global reserves of crude oil are discovered and the old are depleted. Certainly, the diesel engine will continue to seek parity with the SI engine in terms of specific power and sociability, too, even as that engine continues to improve.

Injection Pressure Trends

Today, an informal survey of available fuel injection equipment (FIE) suggests that some systems have high-pressure capability but perhaps are weak in the precision of control; for others, the reverse is true. Future "best-in-class" fuel systems are likely to excel in both areas. Injection pressure, whether considered in terms of peak pressure or mean effective pressure, is the fundamental parameter that affects the rate of fuel-air mixing prior to combustion. As the use of nitrogen oxide (NOx) reducing diluents such as exhaust gas recirculation (EGR) increases, so the need for greater mixing energy increases. To maintain current particulate matter (PM) emissions levels and certainly to improve on them, continued growth in pressure requirement for DI engines can be projected in the future. The actual value will depend on the class of engine, the emissions standard being considered, and the combustion strategy chosen. But as an example, heavy-duty engines in the 2007 to 2010 timeframe are likely to be released with peak injection pressures approaching 2500 bar.

Pressures of this magnitude have not been discussed until recently with respect to common rail injection systems. However, in early 2003, Bosch announced that it was developing an intensified system with 2500-bar potential for heavy-duty engines. This system, referred to as the amplifier piston common rail system (APCRS) [Ref. 7.1], represents one logical approach by which common rail systems can move beyond the apparent 2000-bar barrier that is imposed by the single-stage high-pressure pump used in light-duty systems. In this new system, a large-displacement multicylinder pump pressurizes the fuel rail up to a modest maximum of perhaps 750 bar, and this pressure then is available to the injector via the normal jumper pipes. Internal to the injector are two control valves: one is a nozzle control valve as for any common rail injector, and the other controls the amplifier piston. For light-load conditions, injection is made from the fuel rail. But as load increases and high-pressure injections are required, the second control valve is energized, which brings the intensifier into play. Through its area-ratio amplification, the injection pressure now becomes that much greater than the rail pressure. With this system, low-pressure injections are available for pilot and late post injections, and amplified high pressure may be invoked for the main injection. Given this technology, it can be seen that common rail systems are assuming many of the attributes of two-valve electronic unit injector (EUI) systems.

It seems likely that most, if not all, vehicles in the near future will be fitted with exhaust particulate traps of some type to comply with emissions legislation, whether written on the basis of particulate mass (grams per kilowatt hour [g/kW-h], or grams per mile [g/mile]) or particle size and count. This might suggest that engine-out particulate levels could be relaxed, and with them the requirement for such high peak injection pressures. Surely, the argument goes, money saved on the FIE could be applied to the exhaust aftertreatment. But even if that were true in some individual cases, the underlying trend remains upward. In part, this is because boost pressures are also rising, and this translates into higher air density at top dead center (TDC), for which the antidote is higher injection pressure. Additionally, there is a trend toward engine downsizing, in which a smaller engine operating at higher brake mean effective pressure (BMEP) is substituted for a larger engine. In turn, high BMEP is achieved through high boost pressures, short injection durations, and maximum air utilization, all of which demand higher injection pressure, particularly when EGR or other diluent is used. A further underlying trend is that in-cylinder emissions control is currently perceived as being more cost effective than exhaust aftertreatment; thus, combustion improvement for minimum engine-out emissions typically receives highest priority.

Control Trends

As noted, the term "controllability" embraces many aspects of fuel system functionality. In this context, we are referencing those aspects that enable the combustion heat release to more closely match the optimum profile, and the parameters such as shot-to-shot and injector-to-injector consistency that affect the noise, vibration, and harshness (NVH) performance. In turn, these aspects affect emissions performance and customer perception of the product. The more flexible, responsive, and consistent the system is, the closer the calibrator can approach his targets, and the greater will be the customer satisfaction. Again, continued demand for ever-better precision and controllability can be assumed, the better to respond to combustion feedback signals or to operate closer to a desired operating point. The role to be played by neural networks and so-called virtual sensors is unclear at this stage, but it can be assumed that R&D resources will be directed to these technologies because they appear to hold promise for system simplification, cost reduction, and enhanced diagnostics.

Integration Trends

Technological development in other areas of the internal combustion engine (ICE) will impact the FIE; one such example is the anticipated application of "camless" or hydraulic lost-motion valvetrains [Ref. 7.2]. Additionally, once valve actuation is freed from the yoke of a close and constrained geometric relationship to a camshaft, the opportunity to embrace radially disposed valves is likely to be a choice selected for certain applications. One driver for this move is that of competitive advantage, through the chance to accommodate larger valves for better breathing than is possible with vertical valves. This can be done with careful design, without incurring a dead-volume crevice-air penalty. The enabler, of course, is that the electrohydraulic valve actuators necessary for camless actuation do not have the spatial constraints that cams and rockers do. Even small compound valve angles open significantly more space in the center of the head where the injector is located, large ports notwithstanding. This bonus space may then be used for a larger and more robust injector or, more likely, for better cooling of the injector and nozzle. Both features are likely to be welcomed by the FIE manufacturer and will be necessary as brake mean effective pressure (BMEP) levels rise [Ref. 7.3].

A secondary impact of variable valve timing (VVT) on diesel common rail FIE is the likelihood that the hydraulic pump that is necessary for the hydraulic valve actuation circuit will be integrated with the high-pressure fuel pump. This could take the form of either an intermediate pressure stage if fuel is the hydraulic medium or as a separate pump if a different fluid is used. This dual use will have implications on the engine accessory drive for the pump.

Combustion Trends

Premixed combustion is another technological area that will pose challenges to the FIE manufacturer. Under this umbrella fall the technologies variously known as homogeneous charge compression ignition (HCCI), partially premixed compression ignition (PPCI), premixed charge compression ignition (PCCI), and low-temperature combustion (LTC). The drive for a way to meet emissions standards at the lowest possible cost makes these technologies attractive to the engine OEM. But the lack of maturity of these technologies, particularly with respect to control issues, means that they must initially play a subservient role to the conventional combustion mode [Ref. 7.4]. The fundamental problem is that diesel fuel injection systems have been developed and optimized over the years to support a late-injection diffusion-controlled combustion system, whereas by definition, HCCI requires homogeneous fuel-air preparation. Therefore, the system modifications are likely to center on the nozzle because this has responsibility for presentation of the fuel to the air. The thrust of recent diesel-based direct injection HCCI combustion research has been to seek rapid fuel-air mixing with a multi-orifice nozzle, taking care to avoid wall wetting or impingement by the sprays. Injection may take place well before piston TDC, when the in-cylinder air density is relatively low and over-penetration of the spray is a distinct probability. Conversely, the spray targeting optimized for normal (late) injection at TDC has a relatively short distance to reach the cylinder wall. This situation is likely to result in a requirement for a mode-switching nozzle, in which nozzle flow area, orifice number, and spray angle can be selected on demand among two or more options. This will not be easy to achieve, but it is likely to be attempted with variable area nozzle (VAN) designs presently under development.

An alternative geometry that is receiving significant attention at present is that of smaller included nozzle spray angles, as exemplified by the Institut Français du Pétrole (IFP) narrow-angle direct injection combustion system [Ref. 7.5]. Whereas the conventional DI combustion system typically finds that the optimum nozzle spray cone angle is in the 145- to 155-degree range, narrow-cone-angle sprays are in the 70- to 90-degree range that sometimes requires a modified combustion bowl in the piston crown. Narrow-angle sprays confer at least two advantages over conventional spray angles. First, under early premixed injection conditions, there is less likelihood of fuel impingement on the cylinder wall. Second, the combustion system should be much less sensitive to piston position than is the case for traditional geometries where the reentrant lip quickly interferes with the spray plume as the piston descends after TDC. This is illustrated in Figure 7.1. Note that to obtain the reversed toroidal squish air motion, narrow-angle spray combustion systems generally require lower swirl velocities, too, which can compromise high-load performance.

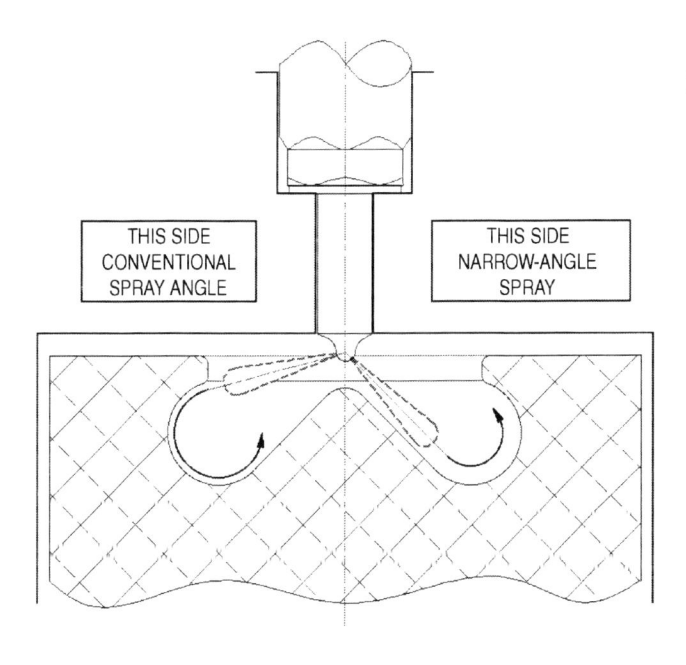

Figure 7.1 (Left) A conventional wide-angle spray pattern versus (right) a narrow-angle spray pattern, showing the associated air motion.

Common Rail Outlook

Today, the light-duty sector has adopted both common rail and, to a lesser extent, EUI solutions. The former is for the bulk of the market, and the latter is being used for high specific power ratings. However, in the heavy-duty sector, the situation is effectively reversed with respect to market share. Beyond 2000 bar, it becomes increasingly difficult to generate the rail pressure in a single stage; therefore, multistage designs are likely to be introduced, involving a medium-pressure rail with pressure intensification at or near the injector. As for the Bosch APCRS, this requires the added complication of a second actuator and valve in the injector: one to control fueling, and one to invoke the intensifier when high pressure is required. Emulating the ability of two-valve EUI to control injection rate on a shot-by-shot basis is likely to be a development goal. The medium-pressure rail may improve the synergistic position of high-pressure common rail (HPCR) FIE with respect to integration with electrohydraulic valve actuation schemes in the future, because potentially a single fluid can now be used for both systems.

Variable area nozzles will play a key emissions reduction role in the future. Conventional nozzle flow area and geometry are optimized for the rated power condition. At any load condition below that, the flow area is larger than necessary, which has negative implications for spray quality as discussed in Chapter Two.

It has been noted that the ever-improving responsiveness of common rail injectors makes the concept of closed-loop control a realistic goal to pursue. The challenge will be to deliver all of these advanced technologies at a cost that still permits the diesel engine to be competitive with the continuously improving SI engine. Cost reductions will come from continuous efforts at exploiting system integration to both base-engine and the other subsystems such as air management. As FIE complexity and sophistication increase, so too does the system cost; value engineering will exert some control over this, but the economics of high-volume production will probably continue to have the largest impact.

Outlook for Electronic Unit Injectors

The higher-peak-pressure capability of the EUI and electronic unit pump (EUP) systems mean that they are likely to remain the system of choice for highly rated engines. A 2000-bar peak injection pressure is common today, with 2500 bar in near-term prospect. Laboratory combustion studies have shown that heroic pressures of 3000 bar result in very clean combustion, and these results are likely to maintain development focus on continued pressure escalation.

Electronic unit injectors can be expected to follow many of the common rail trends, such as adoption of piezoelectric control actuators, variable area nozzles, and control loop-closing strategies. Current deficiencies of only being able to inject fuel in the crank-angle domain where a cam is present will be overcome, so that nontraditional combustion cycles can be entertained. By combining a medium-pressure rail with the unit injectors, and charging that rail from the injectors, it becomes possible to provide for relatively low-pressure injections for premixed charge preparation, for pilot injections, or for aftertreatment regeneration from the rail, while leaving the cam-driven injector to deliver the high-pressure main and post injections. Ultimately, however, should engines make the move to camless operation, an unwelcome possibility for a cam-based FIE, there will likely be a gradual evolution toward hydraulic electronic unit injector (HEUI) architecture.

Market Outlook for Diesel Fuel Injection Equipment

The diesel engine appears reasonably secure in most markets that it inhabits, and that outlook seems set to prevail for the next twenty years or so. It is particularly true for the industrial, marine, and off-highway markets and for the commercial truck market, too, although there have been some incursions by alternative fuel engines on the grounds of improved emissions performance. However, as global emissions standards tighten, it seems probable that the diesel engine fraternity will rise to the challenge. Providing they do so with an inquiring mind open to all technical possibilities and with the courage to pursue nontraditional solutions, a cost-effective compliant powerplant will result. As an example, contemporary research into HCCI and other low-temperature combustion modes appears likely to improve the position of the diesel engine with respect to exhaust emissions, while converging in certain respects with the gasoline engine.

Of the markets in which the diesel engine has yet to make significant headway, the light-duty automotive and small industrial engine sectors in the NAFTA region are perhaps the most interesting to consider. Numerous market studies have been published in recent years, contrasting the North American market with that of Western Europe, with explanations for the divergence in automotive engine purchasing preference [Ref. 7.6]. General agreement is reached that taxation policy has been a strong driver in Europe, where high fuel prices and typically a small price advantage at the pump for diesel fuel provide an economic incentive for encouraging new-car purchases in favor of diesels. The relative importance placed on greenhouse gases such as CO_2 and their reduction in Europe is a factor in the taxation policies for that market, but it is not strongly echoed in North American policies, although California is leading the nation in that direction.

Many other factors and imponderables at the consumer level influence transport purchase decisions, but the experience in Western Europe has been that the modern turbocharged diesel engine brings a fun-to-drive factor to the car ownership value-equation that often is the deciding factor. In large part, this is due to the impressive torque characteristics of the engine in the normal road-load speed range in which most driving time is spent. Consumer satisfaction with this parameter may be expected to be further enhanced when the advantages of variable timing of inlet valve closing on low-speed torque are made available to diesel engines [Ref. 7.7].

Future Scenarios

We should now make an attempt, however futile, to look into the future and project the possible development path or paths that the diesel engine will take. This is necessary because the future of diesel FIE naturally is closely tied to that of the combustion process, and whither goes that, so goes the FIE industry. This section will bring together many of the factors already discussed here and, in the process, suggest ways in which technology combinations can be assembled that result in engines that meet legislative and consumer requirements.

Drivers

The specifications of current engines in whichever market sector you care to examine have been determined over a considerable period of time under the influence of a number of common drivers. Among these are the following:

- **Legislation:** Exhaust emissions, CO_2, noise, pedestrian safety (hood height)
- **Durability:** Emissions warranty period, life expectancy
- **Costs:** Manufacturing cost, selling price, operating cost, whole life cost
- **Performance:** Specific power, specific weight, cost per horsepower
- **Fuel consumption:** Specific fuel consumption (SFC), distance between refueling, fuel costs
- **Quality:** NVH, refinement, reliability, warranty experience
- **Image:** Green, sporty, high versus low tech, rugged, exclusive

As a result, there is a lot of similarity among engines within a single market sector and particularly in a common geographic region, because the relative weight assigned to these drivers can vary. As an example, and despite attempts at world harmonization, the emissions test cycles are distinctly different among the United States, Europe, and Japan. Thus, although the base engine design may be the same for all regions, subtle differences in build and calibration will exist. These differences can influence where the available R&D dollars should go, which is a fact that can be illustrated by noting that Japan took an early lead in HCCI research, while European manufacturers typically focused on engine sophistication and performance.

With respect to emissions, the U.S. light-duty Tier 2 Bin 5 standards are very challenging, as will be the U.S. heavy-duty 2007/2010 standards. However, the feeling is that the currently controlled emissions sensibly cannot be taken further; therefore, the focus is already beginning to move to regulation of CO_2 [Ref. 7.8]. This awareness is more advanced in Europe, in large part due to its stronger commitment to meeting the Kyoto Protocols on greenhouse gas emissions reduction, but also arguably because the strong relationship between CO_2 and carbon-based fuel consumption reduction is in better alignment with the underlying sociopolitical goals of that populous region.

Because of the strong link between CO_2 reduction and the engine technologies most able to minimize fossil fuel consumption, it would seem that the diesel engine has a secure position in the future until such time as it may be displaced by a lower-cost powerplant of similar or better well to wheels efficiency. Although the future may appear bright for the diesel engine, it is expected to remain under serious challenge from the SI engine, which still has several fuel economy improvement options available to exploit. These include idle shutdown, cylinder disablement, valve event optimization, and nonhomogeneous DI calibrations. Further, the application of exhaust aftertreatment to the diesel engine will degrade its fuel economy advantage *vis-à-vis* the SI engine by an amount perhaps as high as 7% (although normally much less), depending on the hardware and control strategies chosen. This is due to the probable need for regular hydrocarbon (HC) (fuel) dosing into the exhaust system for exothermic regeneration of the PM trap and the NOx adsorber.

Note that where selective catalyst reduction (SCR) is selected as the NOx reduction technology, continuous dosing of an ammonia-based compound such as aqueous urea is required at a level corresponding to approximately 3% of fuel consumption, but this fluid does not count against fuel consumed and thus is beneficial in meeting Corporate Average Fuel Economy (CAFE) standards. If higher rates of reductant dosing can be tolerated, there is the possibility of operation at more advanced injection timings, in which case a further fuel economy gain can be realized. Recognize that although the diesel engine will normally take a fuel consumption penalty as exhaust aftertreatment is added, the SI engine already accepted that hit when it adopted the three-way catalyst requiring homogeneous operation under all operating conditions, where previously it would have been able to operate at a lean air-fuel ratio under road-load conditions.

Strategy Selection

In a manner similar to that already discussed, in which it was projected that diesel fuel injection systems are likely to converge on a system that is neither HPCR nor a conventional unit injector, so it seems probable that in the fullness of time, the Otto and the diesel engines will converge on a cycle that we at this time will call HCCI. Certainly, a considerable amount of HCCI research is being conducted with both gasoline and diesel fuels, and it is possible that the optimum may be a blend found somewhere between those two specifications. A preliminary technology road map can be deduced from the direction of current research and can be articulated at a high level by anticipating that progress will be made from conventional "diffusion" diesel combustion, into mixed-mode premixed charge preparation combustion modes, and ultimately to totally premixed HCCI. This is illustrated in Figure 7.2. Progress toward this same goal, sometimes referred to as controlled auto-ignition (CAI), is being made from the other direction by researchers in the SI engine arena.

Motivation for Taking This Path

Some discussion is necessary about the motivation for making progress in this direction. If we examine the case of a pre-existing diesel engine operating in the conventional diffusion combustion mode, it presently is not conceivable that such an engine would be able to meet Euro V or U.S. Tier 2 Bin 5 requirements without comprehensive PM and NOx aftertreatment, and definitely not for any follow-on standards beyond these. This poses a problem for diesel engine OEMs because the anticipated costs for aftertreatment are very high. The total powertrain cost will now place the diesel engine at an even greater disadvantage relative to the SI engine. In turn, this may depress diesel market penetration, while the new addition of aftertreatment that is a purchased commodity does nothing to improve the bottom line of the OEM anyway. Thus, there is a strong incentive to develop clean combustion technologies that require only minimal aftertreatment, and more particularly those technologies that minimize the use of precious metal group (PMG) materials.

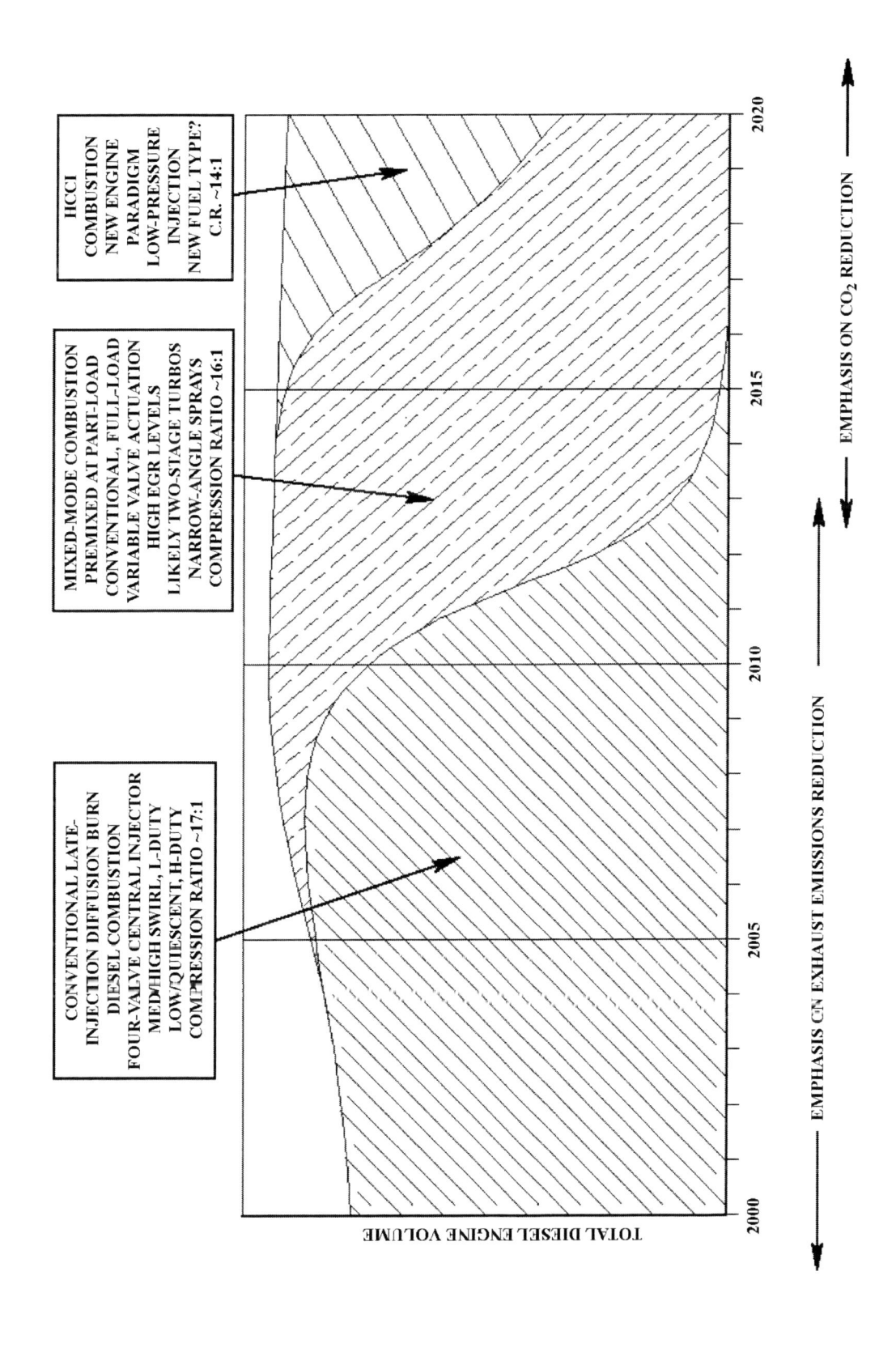

Figure 7.2 A speculative timeline for diesel combustion system evolution.

Now, it so happens that research into HCCI concepts has demonstrated very low engine-out values of NOx and PM, albeit over a restricted range of speed and load, but this finding is pointing the way ahead. Current R&D is largely focused on understanding the chemical kinetics of the auto-ignition process and expanding the island of satisfactory HCCI operation to approach that of a conventional engine.

Unlike the diffusion burn of conventional diesel combustion, HCCI implies a spontaneous auto-ignition of the premixed charge, which, in theory, happens uniformly throughout the combustion chamber when the conditions are right. Whereas with diesel combustion, the bulk air-fuel ratio may be lean, but the local air-fuel ratio at the point of burning is close to stoichiometric and certainly within the flammable range. With HCCI combustion, the bulk air-fuel ratio is lean and nominally homogeneous. Under appropriate conditions of temperature and pressure for the particular fuel properties, the spontaneous, flameless reaction will occur. But because of the well-mixed and overall lean conditions, there are no fuel-rich areas to create particulate precursors, nor fuel concentrations (as in a plume) that generate high temperatures and thus NOx. However, because the cylinder contents are uniform, as in a well-stirred reactor, and therefore tend to reach the reaction initiation state concurrently, the resulting heat release can lead to very high rates of pressure rise in the cylinder, having both structural and NVH implications. Thus, it will be seen that HCCI combustion is unique because it does not involve a spark-induced flame front moving through a stoichiometric homogeneous mixture, nor is it a diffusion flame that exists for as long as fuel and air are provided. It is, in fact, a kinetically controlled spontaneous auto-ignition reaction similar to end-gas knocking in an SI engine, but under very lean conditions.

However, HCCI is not without its problems. As with any premixed charge system, not all the fuel takes part in the combustion process, because some is quenched in out-of-the-way pockets and crevices, with the result that exhaust emissions of HC and CO typically are very high and must be remediated with an oxidation catalyst. This is generally seen as being entirely manageable because such catalysts are mature technology, which is not the case for active PM and lean NOx catalysts. Having said that, combustion temperatures and thus exhaust temperatures under HCCI operation are low, particularly at light load. Therefore, low-temperature oxidation catalyst formulations and/or exhaust port catalysts will be required. The major problem with HCCI is the issue of arranging for the start of reaction (SOR) to occur at the optimum crank-angle timing. There is no singular positive control action available that equates to the spark timing of the Otto cycle nor the start of injection of the diesel cycle. Thus, indirect parameters such as charge temperature, boost pressure, effective compression ratio, and internal/external EGR percentage are employed for this purpose. In addition, neither of the currently available primary fuels, gasoline and diesel, are truly optimum for this combustion system [Ref. 7.9].

Successful HCCI operation over a wide speed and load regime requires flexible and real-time manipulation of the above-mentioned parameters and therefore is fundamentally a controls challenge. The key point, however, is that no mention of the fuel system was made as a controlling influence as it most definitely is for conventional compression ignition (CI) combustion. This is because premixed fuel preparation can potentially be handled by low-cost "gasoline" port or gasoline direct injection (GDI) fuel injectors, thus at one stroke eliminating the expensive high-pressure diesel FIE. Now it remains to be seen if high-pressure direct in-cylinder injection will provide benefit to HCCI combustion, but from this distance, HCCI could be the beginning of the end for conventional diesel FIE.

Having revealed the ultimate denouement for high-pressure diesel injection systems, we now need to chart the road map for how we get from the "here and now" in 2005 to the "there and then" in circa 2015. At this time, HCCI is a very immature technology and very much in the R&D phase. Thus, a reasonable path forward is to take certain elements of the concept and gradually integrate them into current engines. This will provide the double benefit of lower emissions from premixed charge preparation over those parts of the speed-load range where it can be made to work in a current engine, and second, some real-world experience of the new technology at a reasonably low risk. This first tentative step has already been taken by some Japanese OEMs and is expected to be followed by most other manufacturers in due course. Therefore, let us examine the likely path in greater detail, reviewing possible options along the way, and try to identify some logical options.

Current Conventional "Diffusion" Combustion

Very general detail of current engines has been covered already in this book, and specific information is available from the manufacturers; however, a brief recap is appropriate. Direct injection combustion systems are now the dominant type, having taken over from indirect injection (IDI) systems quite recently in the light-duty automotive field, in large part due to the availability of the common rail injection system and the functionality that it brings. In such systems, the injector sits central and vertical in the cylinder head and incorporates a multi-orifice nozzle, having a wide-angle spray trajectory that cooperates with a carefully profiled but relatively shallow combustion bowl in the piston crown. Fuel is injected in one or a multiple of discrete events, but all within the crank-angle vicinity of TDC following the compression stroke. Combustion commences after a very short ignition delay period, and subsequent injected fuel parcels feed that initial conflagration at an intensity that is a function of the injection rate and the fuel-air mixing rate. The adiabatic flame temperature in diffusion combustion can easily exceed the 1900 degrees Kelvin at which NOx is formed. Combustion continues after the last fuel parcel has been injected, until it is all consumed, but the mixing rate declines toward the end of the combustion event. The result is that some liquid fuel pyrolyzes to carbon before it has found the necessary oxygen to burn, hence producing smoke and PM. Diesel combustion is efficient, resulting in excellent SFC and low emissions of HC and CO. But even with the benefit of modern fuel injection systems, further in-cylinder reductions of NOx and PM nevertheless will be difficult to achieve.

As discussed previously, to address the emissions challenge of 2007/2010, high levels of EGR will be utilized—throughout the speed-load envelope in the case of heavy-duty engines, but only in the low- and mid-speed ranges covered by the drive cycle, in the case of light-duty vehicles [Ref. 7.10]. To maintain present power ratings, this strategy will require high boost pressures to make up the oxygen displaced by diluent, and this, in turn, will require sophisticated turbomachinery and high injection pressures to penetrate the dense air. Therefore, the FIE for this generation of engine will be similar to previous marks but will have been up-rated for robustness at circa 1800 bar for light-duty engines and 2500 bar for heavy-duty engines, albeit with ongoing incremental improvements in control responsiveness. A corollary to the higher-pressure capability and control refinement, is, of course, higher cost for the FIE. However, despite resistance that can be anticipated from the engine OEMs, no other cost-effective solution is expected to appear in the near term to change this scenario.

If a conventional diffusion combustion strategy is assumed, then the emissions standards promulgated for the 2010 timeframe are fully expected to require a comprehensive suite of exhaust aftertreatment: oxidation catalyst, PM trap, and NOx adsorber or SCR catalyst. This solution could be considered a benchmark against which all other strategies will be judged, because nominally, little physical change would be made to the engine, and this therefore might be thought of as the "safe" option. However, the cost of a full complement of aftertreatment has variously been estimated between 40% and 90% of the base engine cost, depending on the system strategy selected and the quantity of PMG materials used. It goes without saying that sudden cost increases of this magnitude are not readily accepted in the marketplace and that there will be strong incentive and motivation to seek lower-cost solutions.

Lower-Cost Alternatives

What are the opportunities to reduce cost? An initial examination of the high-cost modules does not appear promising: the reciprocating engine is a mature technology from which little meaningful cost should remain to be extracted. The common rail FIE and turbo air handling systems are relatively new technologies that are rapidly evolving and could be considered moving targets, and the aftertreatment is so new, with much of it being first generation, that the cost is what it is. For the radical reduction in cost that is needed, a more fundamental approach must be taken, and the engine as a whole must be reexamined.

Continuing the examination of a "diffusion combustion only" strategy, a revisitation of IDI or swirl chamber combustion systems must be an option, however "against the grain" that may seem. It is at least worthy of consideration, because it is too easy to slip into the black-or-white mindset of IDI versus DI, Ricardo Comet versus AVL direct injection chamber, when there is, in fact, a continuum between them. Somewhere on that continuum, a new "sweet spot" may exist. As noted, the move from legacy IDI combustion systems was made due to the improved fuel consumption and higher specific power available from DI engines. But as pointed out in the trilogy of SAE papers [Ref. 7.11], the limiting problems with divided-chamber engines have never been addressed in a methodical way, and there remains, in fact, plenty of development potential left in such engines, providing a new combustion chamber is accepted. As an example, the S.E.M.T. Pielstick PA4V200 engine is a currently available divided-chamber engine intended for rail traction applications that has a novel variable geometry prechamber. The piston has a "poker-pin" that enters the throat of the prechamber as TDC is approached, but leaves it unobstructed for the bulk of the cycle, so that pumping losses are minimized [Ref. 7.12].

Traditionally, exhaust emissions from IDI engines are generally lower than for comparable DI engines, and this is borne out by the Pielstick engine, which claims half the NOx emissions for this variable geometry chamber version relative to its DI siblings. Additionally, divided-chamber combustion systems must be more suited to the three-stage "lean-rich-lean" strategies that are used by gas turbine engines to achieve low emissions. Given this reality, it then becomes sensible to consider if a new balance exists between DI and IDI engines once exhaust aftertreatment is applied. Taking the case of an emissions-compliant light-duty DI engine with lean NOx trap (LNT) NOx reduction technology, a real-life fuel consumption penalty of 5% due to the fuel dosed to the catalyst is a reasonable assumption [Ref. 7.13]. If we further assume that a modern low-loss IDI combustion system could get within 5% or 6% in terms of SFC to a DI engine and at which point the smaller LNT dosing penalty is added, we then have an engine that is perhaps 4% worse on CO_2 emissions but is cheaper to manufacture. Traditionally, firing pressures have been lower in IDI engines, and if that observation holds true for a modern incarnation, then a lighter structure becomes possible, with all the knock-on implications that brings. Likewise, lower levels of EGR for the same engine-out NOx emissions might be possible, in which case, lower boost pressures and cheaper turbomachinery could be realistic. A further significant cost reduction would be obtained from the common rail FIE because a higher proportion of the mixing energy will be coming from the air motion. This would allow the injection system to operate at much lower pressures, say, 1200 bar, rather than pushing the envelope at 1800 bar or more in the DI application.

The Buchi Combustion System

An alternative combustion system that fits the description of being on the continuum between conventional IDI and DI is that propounded in the mid-1950s by Dr. Alfred J. Buchi, whose not-inconsiderable claim to fame was the invention of the turbocharger. He developed a combustion system that was essentially a swirl chamber, familiar to us from IDI engines, but in this case, it was in DI format [Ref. 1.3]. A problem with conventional DI engines is that it is difficult to generate high levels of charge-air swirl efficiently; moderate swirl can be generated with helical and directed ports, but at some cost to fuel consumption and nowhere near the swirl levels available in IDI engines. Hence, the fuel-air mixing energy must come, for the most part, from the fuel momentum, implying high injection pressures.

In the Buchi system, the air swirl is generated efficiently because the free-flowing inlet port has a large well-defined helix, and it is truly on-axis to the cylinder, so that the resulting mean swirl level is high. Then, because the combustion bowl is on center, too, the final swirl level at TDC is, through the law of conservation of momentum, at near-IDI levels. All this is achieved without restrictive swirl chamber passages. However, to have a large inlet valve on cylinder axis obviously implies some level of unorthodoxy because the injector can no longer occupy this space. Therefore, to make his system work, Buchi developed an arrangement of concentric telescopic valves, as shown in Figure 7.3. The swirl chamber is formed in a bowl between the piston and the concentric valve heads; therefore, it makes sense for the injector (or injectors) to approach from the side. The chamber has a comparatively low surface-to-volume ratio, one side of which is taken up with the valve heads, so that there is less area potentially exposed to coolant than in a conventional engine, IDI or DI, thus leading to an expectation of good fuel consumption. Moreover, the thermal loading is entirely symmetrical around the cylinder axis without power-limiting constraints of exhaust valve bridge-cooling that occur in conventional engines.

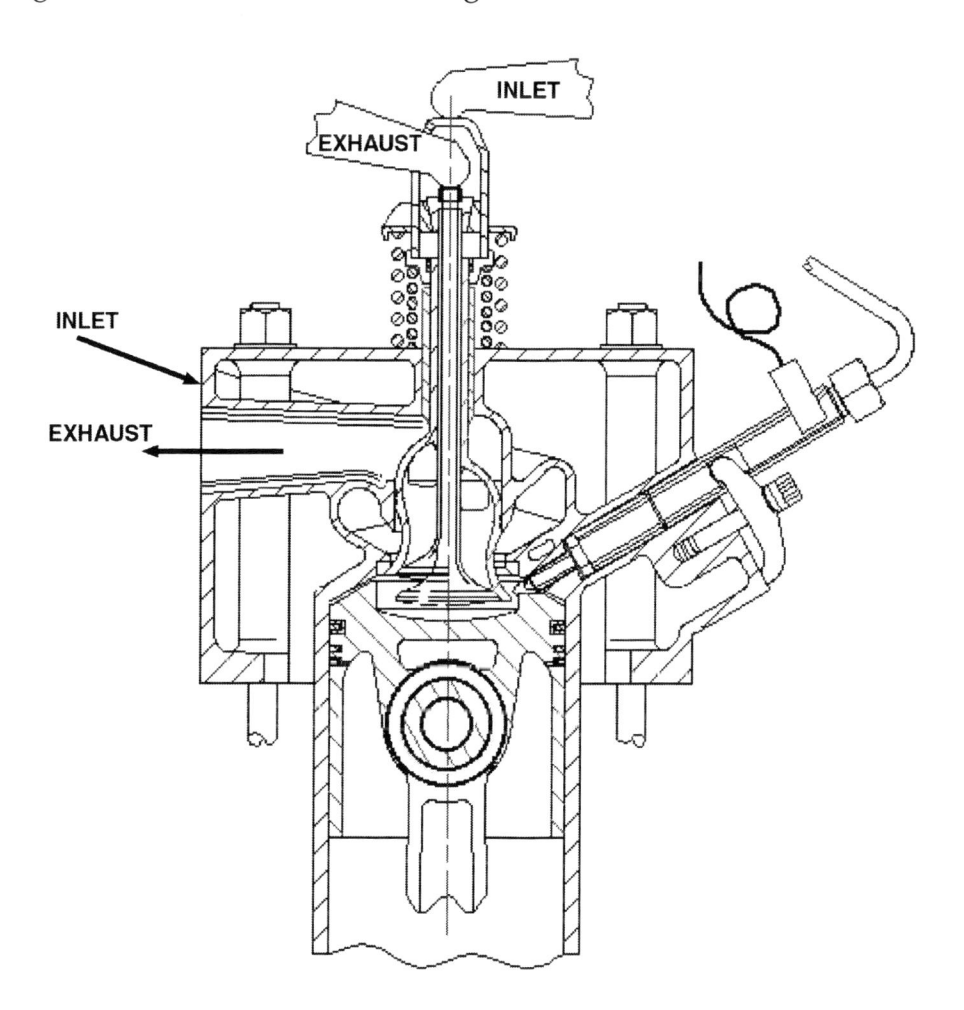

Figure 7.3 A split sectional view of a conceptual Buchi combustion system with concentric telescopic valves, shown in a fixed-head format with common rail injector.

Similar to the proposed IDI engine, the fuel injection system implications are that a relatively low injection pressure, say, 1400 bar, with a single hole nozzle would suffice. However, for minimum emissions, all modern combustion systems have evolved to multiple small holes, which at first glance would suggest that this single-hole/single-injector system would be uncompetitive. However, there are two mitigating circumstances. First, conventional engines with central injectors have a relatively restricted free-plume length (from nozzle tip to combustion bowl wall), which can be a mixing limitation, whereas in approaching from the side, the Buchi system has a much longer free-plume length probably equal to two conventional plumes. Moreover, it is possible to use a long-stem pintle nozzle that gives a hollow-cone spray plume, which accelerates the mixing rate relative to a single plume. Second, there is the opportunity to add another injector on the far side of the chamber, which then enables staged injection strategies not available from a single injector. If the two nozzles are given different flow rates, then the effect of a VAN may be achieved without the complexity of those devices. Indeed, the logical FIE for this engine would be two low-cost solenoid common rail injectors, perhaps operating at two different rail pressures, rather than the one high-cost piezoelectric injector in a competitive conventional engine.

No emissions data exist for the Buchi combustion system, but it likely falls between IDI and DI levels. Likewise, manufacturing costs are unknown, but the straightforward nature of the cylinder head cores and the simple on-axis machining of the bore and valve seat suggest that the design lends itself to a low-cost fixed-head embodiment. BMEP levels will not be as high as could be attained from the best conventional open chambers, but they would likely be on par with the average of North American automotive engines. Certainly, satisfactory development of the valvetrain for modern engine speeds would be a challenge, but Buchi offered several system proposals in his multiple patents of the concept [Ref. 7.14], and lighter-weight materials and hollow valves are more readily available today. Finally, because the valves fit within the combustion bowl and thus there is no possibility of piston-to-valve collisions, there is also no restriction on valve overlap at TDC. This fact makes the Buchi system eminently suitable for camless and variable valve actuation (VVA) engines because no malfunction of the actuation mechanism will result in engine damage. Additionally, flexibility of overlap can be used for chamber cooling and improved scavenging, which may make this combustion system particularly suited to four-stroke/two-stroke mode-switching engines.

Summary

Two alternative concepts have been suggested here for conventional diffusion combustion systems that might be lower in engine-out emissions than the traditional DI geometry and perhaps lower in total powertrain cost. Both are somewhat unconventional, but the development work involved to achieve the base-engine cost and emissions reduction potential is all in-house engine OEM core competencies, which should be more palatable than putting total reliance on outsourced aftertreatment development. When contemplating unorthodox concepts, it is often helpful to consider the situation in reverse and imagine that the new proposal is, in fact, the convention, and to examine what one's reaction would be if someone were to propose that which is actually today's solution. What negatives would one perceive in it?

Partially Premixed and Mixed-Mode Combustion

Research conducted primarily in Japan by Nissan, Toyota, Mitsubishi, and others has demonstrated the potential and feasibility of low engine-out NOx and PM emissions through the use of fuel premixed with the intake charge during the intake stroke or early during the compression stroke. As discussed in Chapter Two, this work has resulted in the Nissan modulated kinetic (MK) engine and the Toyota UNIBUS combustion system, among others. In these concepts, an attempt to break the traditional NOx/PM trade-off has been made by premixing as much of the charge as possible, so that part of the fuel at least has had plenty of time to become well mixed. In other words, the time allocated for mixing of the charge becomes a controlled variable, perhaps occurring over 180 crank degrees or more. This is in stark contrast to conventional diesel combustion, where 30 degrees is a typical value. Figure 7.4 shows an indicator diagram from an optical engine in which the fuel is introduced early, following which there is a cool preflame reaction and then the bulk of the fuel ignites. In this instance, there is also a final diffusion burn from fuel poorly mixed at the end of injection and from the nozzle sac.

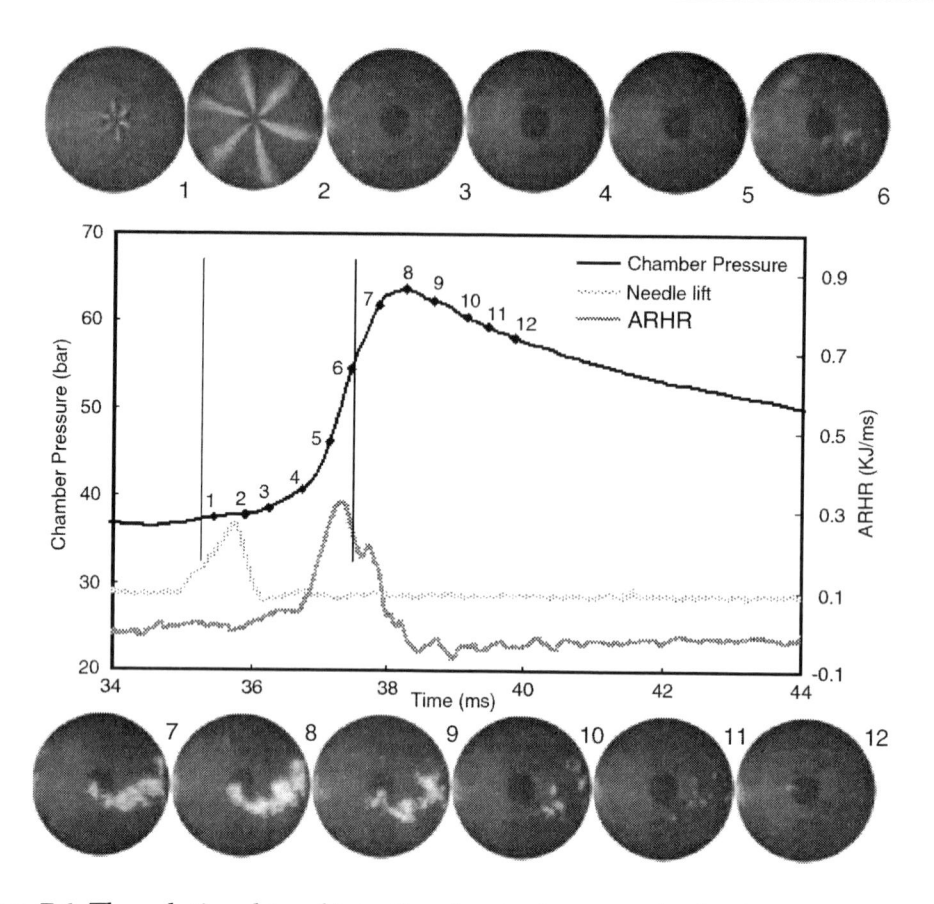

Figure 7.4 The relationship of injection, heat release, and cylinder pressure from a late injection low-temperature low-emissions combustion event.

Simple fumigation of a portion of the fuel charge into the intake of a conventional diesel engine has been studied on many occasions in the past, and the effects are well understood. Typically, an HCCI-like combustion reaction from the fumigated fuel occurs early, before TDC, and this shortens the ignition delay of the main injected fuel quantity, resulting in lower combustion noise and somewhat lower NOx, but higher HC, CO, and SFC numbers [Ref. 7.15]. Although the high HC and CO emissions are to be expected from the premixed fuel charge, the lower fuel efficiency is due mainly to the negative work delivered by the premature heat release.

As it stands, partial fumigation offers insufficient merit to pursue; however, if the reaction could be retarded to occur after TDC, a net benefit might result. This subsequently has been realized through the combination of reduced compression ratio and controlled EGR, both of which effectively postpone the start of reaction. By this means, the centroid of heat release can be moved closer to the optimum timing and allowing the NOx benefit to be taken without the otherwise severe SFC penalty.

As we examine other possible combustion strategies, it will be seen that the ability to influence the time or crank angle of start of combustion becomes an increasingly important ability. Gasoline engines also utilize a prepared mixture, but because the fuel has been selected to have high resistance to self-ignition (as expressed by the octane number), it does not spontaneously ignite when under compression in the cylinder. Rather, it waits for the passage of the spark before doing so. However, diesel fuel has been selected for its ease of self-ignition (as expressed by the cetane number), and when used as a prepared homogeneous mixture in a conventional diesel engine, it is liable to combust earlier than desired. Reducing the compression ratio is an effective countermeasure, as is cooled EGR.

Of course, the problem with lowering the compression ratio is that the cold-start performance (measured as "time to first fire") will be degraded, and lowering the ratio enough to significantly retard ignition of the premixed charge will make cold starts completely unacceptable. Thus, a practical embodiment of a PPCI combustion system is likely to incorporate a VVA valvetrain of some type, so that the effective compression ratio can be controlled. The Caterpillar ACERT™ engines are a representative example of this concept. Recall that in a conventional fixed valve-event engine, the inlet valve closes when the piston is approximately 25% of the way up on the compression stroke. Thus, under cranking conditions, the effective compression ratio is about one ratio lower than the nominal (geometric) compression ratio. Variable valve actuation mechanisms that are particularly suited for diesel engines allow the inlet valve to close anywhere between bottom dead center (BDC) (for cold starting) and perhaps 50 degrees before TDC, so that the effective compression ratio can be changed to lower values on a cycle-by-cycle basis to suit the instantaneous engine conditions.

Exhaust gas recirculation is another parameter that can influence the start of combustion, either by acting as a diluent to delay the initial reaction or, in the case of uncooled "internal" EGR, to accelerate the reaction. Because internal EGR is most conveniently managed through modified valve overlap events and, in turn, these are practical only with a flexible VVA system, it can be seen that combustion strategies that involve either partial- or full-charge premixing are likely to adopt VVA. Once this system is in place, the incentive to employ Miller cycle operation that is thus enabled will be strong, so that SFC and NOx benefits may be exploited. But Miller cycle requires a high level of external compression and aftercooling of the charge, followed by a low internal compression achieved through late intake valve closing. Following combustion, relatively high thermal efficiency is obtained, because the normal expansion stroke is greater than the compression work. For such engines, two-stage turbocharging is usually required, preferably with intercooling between compression stages.

It will now be appreciated that the future diesel engine is likely to become increasingly sophisticated and therefore expensive. Such sophistication is probably not warranted or necessary if the previously mentioned benchmark strategy (conventional diffusion combustion plus comprehensive aftertreatment) is adopted. But for those engine platforms that do go premixed and, in so doing, incorporate flexible injection systems, VVA, and high-pressure-ratio turbomachinery, cost reduction opportunities through value engineering are bound to be sought. At this point, the potential synergy between the fuel injection system and the hydraulically actuated variable valve control system comes into focus, and ways in which they can be integrated as a single unit into a cylinder head module will emerge.

A key aspect of homogeneous combustion is that the auto-ignition heat release reaction can occur over a very wide range of lean fuel-air ratios—from very lean right up to the flammable limit of the fuel being used. However, once that critical fuel-air ratio is reached, the spontaneous cool-flame reaction is replaced by a normal stoichiometric combustion process in which NOx and PM are again formed. This fact limits the maximum BMEP under which true HCCI combustion is possible. Therefore, to extend the range, boost pressure must be increased so that higher BMEPs may be reached before the limiting fuel-air ratio is approached. This strategy can be taken only so far because at some point, engine structural limits will intervene.

Engine Design for Premixed Combustion

Accepting that premixed combustion strategies will require VVA valvetrains (with which valve event timings can be precisely controlled on a cycle-by-cycle and cylinder-by-cylinder basis) and high levels of EGR (either cooled external or uncooled internal), what impact will this have on the engine design, and are there implications for the FIE? Conventional four-valve engines permit very restricted valve overlap due to the minimal clearance necessary between the piston crown and valves at TDC to minimize the dead air pockets. Valve heads that are recessed into the fire-deck or pockets on the piston crown are possible, but they are counterproductive, particularly in small high-swirl cylinder sizes. However, this is not the case on large-bore engines with quiescent combustion systems. This constraint may limit the effectiveness of VVA for HCCI control, not for effective compression ratio control but for combustion chamber temperature control because effective chamber cooling from gas-exchange scavenging is not possible. This could encourage engine designers to look at nontraditional chambers that either do not need valve pockets (e.g., Buchi) or other designs that are immune to the swirl-killing impact of pockets.

A goal of combustion development in this area will be to extend the island of premixed operation across more of the speed-load map. All the while the area of premixed operation remains smaller than the conventional combustion area, the chamber design is likely to remain close to the traditional geometry. But as soon as the premixed operating area assumes dominance, chamber designs can be expected to be optimized for this type of combustion rather than for conventional. No conclusion can be drawn from this analysis as to the likely impact on FIE, because Buchi and the divided-chamber engines may operate with relatively low-pressure FIE, whereas engines that remain with low-swirl chambers will likely require high-pressure FIE. However, mode-switching nozzles will be required that can modulate spray pattern on demand between the two combustion modes. In the case of the Buchi system with two injectors, one could be optimized for "down the bore" early premixed injection, and the other for conventional late injection.

It has already been pointed out that under HCCI operation, the auto-ignition characteristic of the combustion process tends to cause the whole charge to react at once, or very nearly so. This results in a cylinder pressure spike that is high relative to SI combustion in which heat release is limited by flame speed across the chamber or, for that matter, diesel combustion in which heat release is limited by the air-fuel mixing rate. This high-pressure spike and resulting dp/dt (delta pressure/delta time) can easily exceed engine structural limits, particularly because boost pressure also must be raised to extend the load range, and the resulting NVH is not helpful, particularly in light-duty vehicles if the goal is to replace an existing gasoline engine. Ways in which this pressure spike can be mitigated will be sought by the engine designer, and again, the divided-chamber engine may offer a solution if similar but different conditions can be created in the main chamber relative to the antechamber. By definition in a homogeneous charge, the air-fuel ratio and the EGR level should be identical in both chambers. But if it can be arranged for the metal temperatures to be different between them (as is usually the case anyway), then the timing of the combustion reaction may also be different, so that two spikes of lesser magnitude occur, rather than one big spike. As before, the implication for FIE is that lower-pressure, lower-cost FIE might be suitable.

Full-Time HCCI Combustion

Given our current level of technology, a practical engine that can run over a broad speed and load range under true HCCI operation on diesel fuel is difficult to imagine. For the purposes of this discussion, we are assuming that the control difficulties will be overcome, particularly that of gaining satisfactory control over the start of reaction timing so that heat release occurs optimally after TDC. High BMEP requires high boost pressures that imply two-stage turbocharging, a form of which is presently used, for example, on the BMW 535d six-cylinder diesel engine, plus flexible valve actuation for effective compression ratio control and internal EGR for the start of combustion control. But even with a low geometric compression ratio, the engine is still liable to be cylinder-pressure limited before rated power is reached.

At present, this would appear to be a stumbling block for an engine that is intended to operate completely in the low-NOx and PM emissions HCCI regime, without transitioning to diffusion combustion at high loads. Additionally, neither gasoline nor diesel fuel is an ideal fuel for HCCI operation, and a blend of existing fuels or a new fuel may be required to fully exploit the potential of this combustion process, as noted. Because current and very likely future diesel FIE systems depend on the inherent lubricity of the fuel for satisfactory tribological survival, the properties of the ultimately selected optimum fuel will influence the future of the FIE industry and its products.

Without doubt, strenuous efforts by research organizations will be directed to solving the practical problems that inhibit the introduction of HCCI engines. This is because the reduced demand for exhaust aftertreatment that the combustion process has demonstrated holds the prospect of a cost-effective engine that also provides diesel engine levels of CO_2 emissions. Mainstream research in this area will naturally be based on existing engine architectures that have been modified to incorporate the necessary features mentioned above, plus other devices that will have been found to be required. Early implementations for sale to the public may appear in conjunction with infinitely variable transmissions (e.g., Torotrak IVT) and hybrid powertrains, because they can be more forgiving of engine shortcomings in terms of drivability and engine speed range. Furthermore, the on-cost of the hybrid motors and controls will be partially offset by the reduced set of aftertreatment.

For a full-time HCCI engine concept then, the fuel system objective will be to provide a uniformly mixed fuel-air charge across the load range in whatever format, port injection or direct in-cylinder injection, best supports the combustion and control system. This task must be accomplished between extremes of ambient temperature, barometric pressure, and engine speed, and it may be that to cover the map, some concepts may require both port and direct injection systems. Dual injection systems on one engine sounds improbable, but if the results justify the means, it could happen. The desire would be to fall back to a single system, but it would probably be a relatively low-pressure injection system, much closer in concept to a GDI gasoline fuel system than to a diesel injection system and therefore much cheaper than we are presently used to for the latter (Figure 7.5).

Of course, radical change to the automotive landscape brought about by the introduction of HCCI engines may not affect only the diesel FIE industry. Today's conventional engine is a relatively robust, high-BMEP, high-mass, low-firing-frequency powerplant, and it has developed that way because high specific torque has been a desirable attribute. But as we have seen, it is problematic to operate an HCCI engine at high BMEP, so perhaps there are other engine concepts better suited to low-emissions, low-BMEP engines? Well, as it happens, there are.

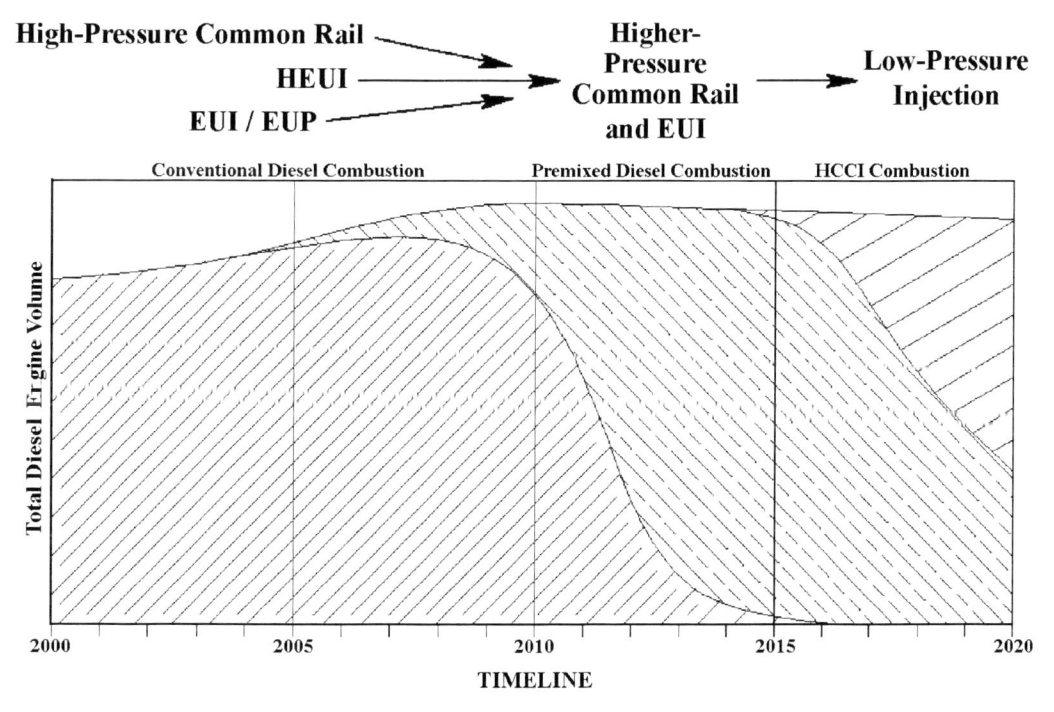

Figure 7.5 A projected scenario for evolution of diesel FIE following the combustion trends.

What is needed is a relatively lightweight structure in which many moderate BMEP combustion events are delivered at high firing frequency, and such engines have been designed, an example being shown in Figure 7.6. This is a four-cylinder opposed-piston two-stroke diesel engine with cylinders arranged in a barrel configuration, in which two opposite pairs of cylinders fire together and the reciprocating motion is translated into rotational through the medium of cam-tracks connected to the central driveshaft. In this compact, lightweight arrangement, eight firing events per shaft revolution are delivered—two every 90 degrees. Additionally, the cam-tracks are not constrained to a conventional sinusoidal motion but can be designed to exploit the combustion process with high- and low-velocity sections at appropriate locations.

This engine was not explicitly designed for HCCI operation, and being port-controlled with uniflow scavenge, there is no ready way to control effective compression ratio (CR) (although geometric CR can be varied by axial movement of the cam carriers), and it would be obliged to have direct injection. Nevertheless, the concept has merit and may not be unique in that respect; it is possible that conventional engines that are configured to operate in a four-stroke/two-stroke switching mode may be able to exploit this ability in the HCCI regime.

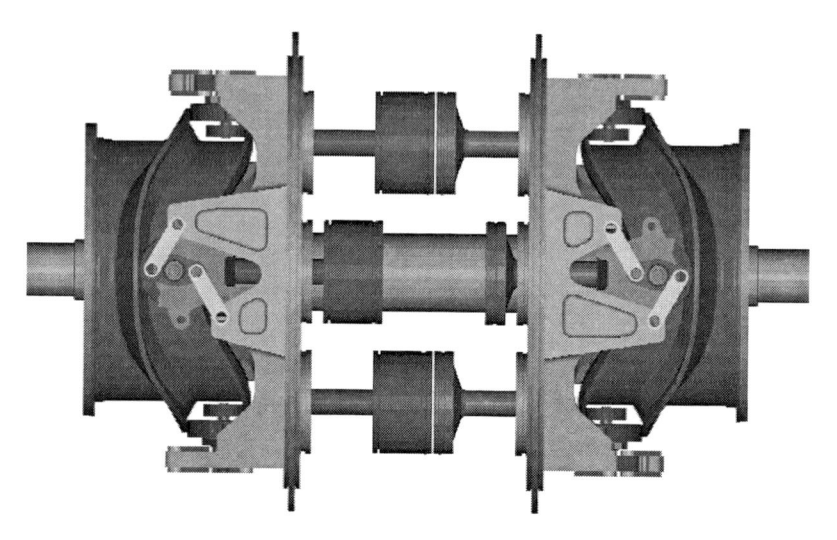

Figure 7.6 Low-BMEP, high-firing-frequency opposed-piston "barrel" engine for HCCI operation.
(Courtesy of Mechanical Innovations and Griffin Motor Works, Inc.)

However, what should be of concern to the traditional engine builders is that engines of this new type, designed to exploit HCCI characteristics, do not have to follow the conventional construction techniques; indeed, with half the BMEP, it would not make sense to do so. Thus, a new engine paradigm may emerge, and with it, new entrants into the market because everyone will be starting from a common basis. Because the timing of this technology introduction, circa 2015, will roughly coincide with rapid growth in the Asian market, particularly China, it could be an opportunity for manufacturers in that region to break out of the hegemonic market stranglehold enjoyed by Western corporations. Interestingly, this potential scenario exactly parallels what has happened recently in the diesel fuel injection market. As reviewed in Ref. 7.16, the investment barriers for outsider entry into the FIE manufacturing community were, for the most part, overwhelmingly high, prior to the advent of common rail technology. But at that point, everyone was starting from a clean sheet of paper. While the barriers were down, Siemens, for example, was able to successfully step into the arena with a differentiated product and become an accredited supplier.

References

Chapter One

1.1 Burman, Paul G. and DeLuca, Frank, *Fuel Injection and Controls*, Simmons-Boardman Publishing Corp., New York, 1962.

1.2 Richards, R.R., *et al.*, "Diesel Engine Emissions Control for the 1990s," SAE Paper No. 880346, Society of Automotive Engineers, Warrendale, PA, 1988.

1.3 CIMAC Congress 1957, paper by Dr. A.J. Buchi, "Four-Cycle Internal Combustion Engines with the Buchi Telescope-Valve System" (copyright CIMAC). International Council on Combustion Engines, Lyoner Str. 18, Frankfurt/Main, Germany.

Chapter Two

2.1 Kimura, S., *et al.*, "Ultra-Clean Combustion Technology Combining a Low Temperature and Premixed Combustion Concept for Meeting Future Emission Standards," SAE Paper No. 2001-01-0200, Society of Automotive Engineers, Warrendale, PA, 2001.

2.2 Akihama, K., *et al.*, "Mechanism of the Smokeless Rich Diesel Combustion by Reducing Temperature," SAE Paper No. 2001-01-0655, Society of Automotive Engineers, Warrendale, PA, 2001.

2.3 Liu, Y., *et al.*, "Optimizing HSDI Diesel Combustion and Emissions Using Multiple Injection Strategies," SAE Paper No. 2005-01-0212, Society of Automotive Engineers, Warrendale, PA, 2005.

2.4 Reitz, R.D., *et al.*, "Mechanism of Atomization of a Liquid Jet," *Phys. Fluid*, 25 (10):1730–1742.

2.5   Soteriou, C., *et al.*, "Further Studies of Cavitation and Atomization in Diesel Injection," SAE Paper No. 1999-01-1486, Society of Automotive Engineers, Warrendale, PA, 1999.

2.6   Qin, J.R., *et al.*, "Transient Cavitating Flow Simulations Inside a 2-D VCO Nozzle Using the Space-Time CE/SE Method," SAE Paper No. 2001-01-1983, Society of Automotive Engineers, Warrendale, PA, 2001.

2.7   Hasegawa, T., *et al.*, "Injection Characteristics and Spray Features of the Variable Orifice Nozzle (VON) for Direct Injection Diesel Engines," SAE Paper No. 980807, Society of Automotive Engineers, Warrendale, PA, 1998.

2.8   Saito, Akinori, *et al.*, Japanese Patent No. JP2000337226, "Atomizing Pattern Variable Fuel Injection Nozzle," publication date December 5, 2000.

Chapter Three

3.1   Russell, M.F., *et al.*, "More Torque, Less Emissions and Less Noise," SAE Paper No. 2000-01-0942, Society of Automotive Engineers, Warrendale, PA, 2000.

3.2   Ohishi, Kazutaka, *et al.*, "The New Common Rail Fuel System for the Duramax 6600 V8 Diesel Engine," SAE Paper No. 2001-01-2704, Society of Automotive Engineers, Warrendale, PA, 2001.

3.3   Guerrassi, Noureddine, *et al.*, "Latest Developments in Diesel Common Rail Technology to Meet Future Demands," SIA International Congress, Lyon, Société des Ingénieurs de l'Automobile, May 2004.

3.4   Morgan, R.R., *et al.*, "A Premium Heavy Duty Engine Concept for 2005 and Beyond," SAE Paper No. 1999-01-0831, Society of Automotive Engineers, Warrendale, PA, 1999.

3.5   Meyer, Steffen, *et al.*, "Flexible Piezo Common Rail System with Direct Needle Control," *MTZ Worldwide*, 02/2002, Vol. 63.

3.6 Kiss, Tibor, *et al.*, "Analytical Comparison of 2 and 3 Way Digital Valves for Direct Needle Control Injectors," SAE Paper No. 2004-01-0032, Society of Automotive Engineers, Warrendale, PA, 2004.

3.7 Cole, C., *et al.*, "Application of Digital Valve Technology to Diesel Fuel Injection," SAE Paper No. 1999-01-0196, Society of Automotive Engineers, Warrendale, PA, 1999.

3.8 Warga, J., *et al.*, "The Future of Piezo Technology," ATA Paper No. 045004, Associazione Tecnica dell'Automobile, 2004.

3.9 Yannick, Raynaud, *et al.*, "Application of Adaptive Online DoE Techniques for Engine ECU Calibration," I-MechE Conference on Statistics and Analytical Methods in Automotive Engineering, Institution of Mechanical Engineers, London, U.K., 2002.

Chapter Four

4.1 Cross, R.K., *et al.*, "Electronic Fuel Injection Equipment for Controlled Combustion in Diesel Engines," SAE Paper No. 810258, Society of Automotive Engineers, Warrendale, PA, 1981.

4.2 Stockner, A.R., *et al.*, "Development of the HEUI Fuel System—Integration of Design, Simulation, Test, and Manufacturing," SAE Paper No. 930270, Society of Automotive Engineers, Warrendale, PA, 1993.

4.3 Coldren, D.R., *et al.*, "Hydraulic Electronic Unit Injector with Rate-Shaping Capability," SAE Paper No. 2003-01-1384, Society of Automotive Engineers, Warrendale, PA, 2003.

Chapter Five

5.1 Soteriou, Celia, *et al.*, "From Concept to End Product—Computer Simulation in the Development of the EUI-200," SAE Paper No. 960866, Society of Automotive Engineers, Warrendale, PA, 1996.

5.2 Greeves, Godfrey, *et al.*, "Advanced Two-Actuator EUI and Emission Reduction for Heavy-Duty Diesel Engines," SAE Paper No. 2003-01-0698, Society of Automotive Engineers, Warrendale, PA, 2003.

Chapter Seven

7.1 Blessing, M., *et al.*, "Analysis of Flow and Cavitation Phenomena in Diesel Injector Nozzles and Its Effect on Spray and Mixture Formation," SAE Paper No. 2003-01-1358, Society of Automotive Engineers, Warrendale, PA, 2003.

7.2 Mardell, J.E., *et al.*, "An Integrated, Full Authority, Electrohydraulic Engine Valve and Diesel Fuel Injection System," SAE Paper No. 880602, Society of Automotive Engineers, Warrendale, PA, 1988.

7.3 Reguiero, Josè F., "ROTULAR TAPPET Valve Trains for Hemispherical Combustion Chambers," SAE Paper No. 960058, Society of Automotive Engineers, Warrendale, PA, 1996.

7.4 Klingbeil, Adam E., "Premixed Diesel Combustion Analysis in a Heavy-Duty Diesel Engine," SAE Paper No. 2003-01-0341, Society of Automotive Engineers, Warrendale, PA, 2003.

7.5 Gatellier, Bertrand, *et al.*, "New Developments of the NADI™ Concept to Improve Operating Range, Exhaust Emissions and Noise," Aachener Kolloquium Fahrzeug- und Motorentechnik, 2004.

7.6 Kliesch, James, *et al.*, "Deliberating Diesel: Environmental, Technical, and Social Factors Affecting Diesel Passenger Prospects in the United States," Report T032, American Council for an Energy-Efficient Economy, Washington, September 2003.

7.7 Fesseler, Harald, *et al.*, "An Electro-Hydraulic 'Lost Motion' System for a 3.0 Liter Diesel Engine," SAE Paper No. 2004-01-3018, Society of Automotive Engineers, Warrendale, PA, 2004.

7.8 Cheetham, Steven B., "European Autos: The Future of Powertrains—Focus on Lower Greenhouse Emissions Moves Fuel Economy to the Fore," Bernstein Research Call, Sanford C. Bernstein & Co., LLC, March 26, 2004.

7.9 Ryan III, Thomas W., "Fuel Requirements for HCCI Operation," JSAE Paper No. 20030353, Society of Automotive Engineers of Japan, 2003.

7.10 Cowland, Chris, *et al.*, "Passenger Vehicle Diesel Engines for the U.S.," SAE Paper No. 2004-01-1452, Society of Automotive Engineers, Warrendale, PA, 2004.

7.11 Regueiro, Josè F., "The Case for New Divided-Chamber Diesel Combustion Systems—Parts One, Two, and Three," SAE Paper Nos. 2001-01-0271, 2001-01-0274, and 2001-01-0278, respectively, Society of Automotive Engineers, Warrendale, PA, 2001.

7.12 http://www.pielstick.com/frRail.html.

7.13 Hoard, John W., *et al.*, "Economic Comparison of LNT Versus Urea SCR for Light Duty Diesel Vehicles in U.S. Market," presentation at Diesel Engine Emission Reduction (DEER) Conference, Coronado, CA, 2004.

7.14 Buchi, J.A., British Patent No. GB949190; also U.S. Patent No. 3055350.

7.15 Simescu, Stefan, *et al.*, "Partial Pre-Mixed Combustion with Cooled and Uncooled EGR in a Heavy Duty Diesel Engine," SAE Paper No. 2002-01-0963, Society of Automotive Engineers, Warrendale, PA, 2002.

7.16 Mardell, John E., "Getting Fun Out of Diesel," Chairman's Address, Institution of Mechanical Engineers, Automotive Division, London, U.K., 2003/2004.

Acronyms and Definitions

APCRS	Amplifier piston common rail system (by Bosch)
Bar	In the context of pressure, 1 bar = 100 kPa = 0.1 MPa = 14.5 lb/in.^{2}
BDC	Bottom dead center
BMEP	Brake mean effective pressure
BSFC	Brake specific fuel consumption
CAFE	Corporate average fuel economy
CAI	Controlled auto-ignition
CARB	California Air Resources Board
Cd	Coefficient of discharge
CI	Compression ignition (the Diesel cycle)
CO	Carbon monoxide
CR	Compression ratio
DI	Direct injection combustion system (an open-chamber system)
DPF	Diesel particulate filter
DSP	Digital signal processor
ECU	Engine control unit
EGR	Exhaust gas recirculation
EUI	Electronic unit injector (a unitized form of high-pressure FIE)
EUP	Electronic unit pump (a close-coupled high-pressure PLN system)
FEA	Finite element analysis
FIE	Fuel injection equipment (the total system)
FMEA	Failure modes and effects analysis

GDI	Gasoline direct injection
GTI	Grooved tip inverted (a current nozzle seat geometry)
HC	Hydrocarbon (emissions)
HCCI	Homogeneous charge compression ignition (an emerging system)
HEUI	Hydraulic (intensified) EUI (a non-cam-driven injection system)
HFRR	High-frequency reciprocating rig (for fuel lubricity evaluation)
HG	Hydro-erosive grinding (an abrasive flow machining process)
HPCR	High (rail) pressure common rail FIE
HPR	High-pressure rotary (a declining PLN system)
ICE	Internal combustion engine
ICR	Intensified common rail (multistage pressure amplification)
ICU	Injection control unit
IDI	Indirect injection combustion system (a divided-chamber system)
In-line pump	A multicylinder PLN system with one pump element per cylinder
IVT	Infinitely variable transmission
LNT	Lean NOx trap (a NOx-adsorbing, HC-regenerated exhaust catalyst)
LTC	Low-temperature combustion (a low-NOx combustion strategy)
Mini-sac nozzle	Nozzle orifii emanating from a small sac below the valve
MK	Modulated kinetic process (by Nissan)
MUI	Mechanical unit injector (pre-EUI injection technology)
NOx	Nitrogen oxides
NTE	Not to exceed
NVH	Noise, vibration, and harshness

OBD	Onboard diagnostics
PCCI	Premixed charge (or controlled) compression ignition
PLN	Pump-line-nozzle system (a traditional FIE architecture)
PM	Particulate matter (emissions)
PMG	Precious metal group
PPCI	Partially premixed compression ignition
ROI	Rate of injection (instantaneous nozzle flow rate)
Rotary pump	A multicylinder distributor injection pump of PLN type
SCR	Selective catalytic reduction (of exhaust NOx)
SFC	Specific fuel comsumption
SI	Spark ignition (the Otto cycle)
SOF	Soluble organic fraction
SOR	Start of reaction (as applied to HCCI kinetic reaction)
TDC	Top dead center
TSI	Two-stage injector
UBHC	Unburned hydrocarbons
UNIBUS	"Uniform bulky" combustion system (by Toyota)
VAN	Variable flow area nozzle (with stepless transition)
VCO nozzle	Valve-covered orifice (a low-HC-emitting nozzle)
VGT	Variable geometry turbocharger
VON	Variable orifice nozzle (in incremental steps)
VVA	Variable valve actuation
VVT	Variable valve timing

About the Authors

Philip Dingle is a Diesel Technology Specialist in the advanced engineering Innovation Center of Delphi Diesel Systems. He received his engineering education in England, and after graduating in 1972, joined the R&D group of Lucas Diesel Systems, where he worked on several advanced engine and fuel system technologies. Transferred to Detroit, Michigan, in the United States in 1975, Mr. Dingle has worked closely with several U.S. diesel engine manufacturers on the development of fuel injection equipment for their engines. In the process, he gained broad experience in achieving performance and emissions targets from both direct injection and indirect injection combustion systems. Mr. Dingle holds twelve U.S. or European patents for fuel system innovation.

Ming-Chia Lai is the Charles DeVlieg Professor of Mechanical Engineering at Wayne State University, where he has taught since 1987. He received his B.S. from National Taiwan University and his M.S. and Ph.D. from Penn State University. He was a Post-Doctoral Fellow at the University of Michigan in 1985 and Massachusetts Institute of Technology in 1986. Dr. Lai has conducted research sponsored by the U.S. Department of Energy, the U.S. Department of Defense, the National Aeronautics and Space Administration, the National Science Foundation, General Motors, Ford, DaimlerChrysler, Detroit Diesel, Delphi, Visteon, Bosch, Eaton, Hyundai, and Ferrari. He has authored or co-authored more than 300 refereed journal and conference papers, with a significant portion in the fuel injection areas.